SpringerBriefs in Materials

More information about this series at http://www.springer.com/series/10111

Shailendra Kumar Singh
Kaushal Kishor · Shanthy Sundaram

Photosynthetic Microorganisms

Mechanism For Carbon Concentration

 Springer

Shailendra Kumar Singh
Shanthy Sundaram
Centre of Biotechnology
Nehru Science Centre
University of Allahabad
Allahabad
Uttar Pradesh
India

Kaushal Kishor
Department of Chemical Engineering
Motilal Nehru National Institute
   of Technology
Allahabad
Uttar Pradesh
India

ISSN 2192-1091   ISSN 2192-1105   (electronic)
ISBN 978-3-319-09122-8   ISBN 978-3-319-09123-5   (eBook)
DOI 10.1007/978-3-319-09123-5

Library of Congress Control Number: 2014947126

Springer Cham Heidelberg New York Dordrecht London

Printed on acid-free paper

Springer is part of Springer Science+Business Media (www.springer.com)

Preface

The brief is very topical as its theme "carbon concentrating mechanism of photosynthetic microorganism" impacts on the lives of all of us. World's rapid industrialization has led to severe interlinked global environmental challenges, such as energy crisis, pollution, and global warming. It is widely recognized that fossil fuel combustion and other emissions resulting from anthrop activity, produces greenhouse gases in which carbon dioxide (CO_2) is the biggest contributor by volume (76 % of CO_2 emissions from Fuel Combustion (2012), International Energy Agency). It has been implicated in the global climate change, and reducing them is a potential solution. In this regard, we must consider photosynthetic microorganism-based carbon mitigation system with the most efficient photosynthetic pathways to reduce the excess atmospheric CO_2 concentrations. The use of such technology will result in balanced O_2 and CO_2 concentrations to mitigate global warming, avoid dangerous climate change, and reduce toxic levels of atmospheric CO_2 through transformation into other forms of carbon. Photosynthetic microorganism such as cyanobacteria and microalgae acquired a special mechanism called "carbon concentrating mechanism (CCM)," which induced in limiting CO_2 environment and increases the CO_2 concentrations actively in the proximity of RuBisCO. Due to the action of CCM, photosynthetic microorganisms have high photosynthetic efficiency than terrestrial C_3 and C_4 plants. CCM is by far the most spectacular physiological process in algal growth and productivity. Due to this fact, the study of CCM has captivated algologist, algae physiologists, algal molecular and cellular biologists, botanists, agriculturalists, crop growers, and most recently algal biofuel researchers around the world. From an esthetic perspective, I thought that it would be wonderful to include many of the remarkable findings on CCM of both important photosynthetic microorganisms in a single inclusive volume.

Book begins with an introduction to the topic, and follows with the reviews on modern carbon sequestration options and proposed measures to stabilize atmospheric CO_2. It then provides state-of-the-art information on CCM of photosynthetic microorganism, as well as also emphasizing its integration with different industry. It helps readers to understand better the interconnection between photosynthesis and CCM and its potential impact on global climate change. This multi-authoritative work by experts also gives insights of metabolic pathways of photosynthetic microorganism, needed to engineer the metabolic pathways of photosynthetic microorganism for enhanced CO_2 fixation. Brief addresses the needs of energy researchers, chemical engineers, fuel and environmental engineers, postgraduate and advanced undergraduate students, and others interested in sustainable energy and environment development. It is also very helpful for Entrepreneurs and Companies planning to start a venture in algae-related industries or businesses with synergistic operations exploring fuel from algae ventures. Brief is highly useful and helps to plan new research and design new economically sustainable viable processes for the production of clean fuels and value added products from algae. Thus, the overall goal of brief is to provide in-depth scientific details on the basic and applied aspects of CCM of both important photosynthetic microorganisms (microalgae and cyanobacteria). Many figures and tables are included in the brief to facilitate understanding and comprehension of the information presented throughout the text. Hundreds of references have been used to prepare this unique collection. The authors welcome reader's comments and suggestions, especially any drawing our attention to errors in the text. The readers can sent their comments and suggestions directly to me by e-mail at shailbiochem@gmail.com.

Allahabad, India   Shailendra Kumar Singh

Acknowledgments

Authors are thankful to the Department of Biotechnology, Ministry of Science and Technology, India for their continued financial support and encouragement through the DBT-RA scheme for our research in carbon concentrating mechanism of cyanobacteria and microalgae. Special thanks go to Dr. Sukrat Sinha, Fulbright Postdoctoral Fellow, Emory Vaccine Center, Atlanta, Georgia (Greater Atlanta Area) for his careful reading and insightful grammatical corrections in the brief. Thanks are also due to the Mr. Md Akhlaqur Rahman, Ms. Kritika Dixit and Ms. Geetika Vajpayee, research scholars at Centre of Biotechnology, University of Allahabad who took the time to read and criticize the brief in order to draw a contextual in-depth overview of carbon concentrating mechanism for readers. Finally, we are also grateful to the University of Allahabad, Allahabad, Uttar Pradesh, India for providing the resources and facility of written materials.

Contents

Chapter 1
Introduction

The capture of light energy from sun and CO_2 from surroundings for transformation, into various reduced carbon forms, of chemical energy is one of the oldest biochemical processes on earth. Early photosynthetic systems, such as in green-purple sulfur and non-sulfur bacteria, are believed to have been anoxygenic, using various molecules such as hydrogen, sulfur, amino, and other organic acids as an electron donor (Lewy 2013). Due to the highly reduced atmosphere, the availability of these molecules was consistent at that time. In contrast, contemporary photosynthesis in plants and most photosynthetic prokaryotes is oxygenic and uses water as an electron donor. Geographical data suggest that about the 2.5 billion years ago of Paleoproterozoic era, organisms such as cyanobacteria that capable of oxygenic photosynthesis have been evolved in parallel to a long-term atmospheric CO_2 reduction and increase in O_2 (Starr et al. 2012). In present-day atmospheres, when temperature, water, and light are not limiting, consequently, the competition between O_2 and CO_2 for the active sites of RuBP carboxylase–oxygenase (RuBisCO) became more and more restrictive to the rate of photosynthesis. Today, majority of the plants on earth are photosynthetically inhibited by a phenomena photorespiration (also known as the oxidative photosynthetic carbon cycle, or C_2 photosynthesis), occurs inside a leaf when CO_2 concentrations become low and O_2 outcompetes with CO_2 for the RuBisCO's active site.

Furthermore, low diffusibility of CO_2 in aqueous environment also poses severe restrictions on photosynthesis (Brueggeman et al. 2012), especially on photosynthetic microorganisms such as microalgae and cyanobacteria. In coping with this situation, many species acquired a special mechanism called "carbon concentrating mechanism (CCM)," induced in limiting CO_2 environment and increases the CO_2 concentrations actively in the proximity of RuBisCO. Many researchers have proved that the high photosynthetic efficiency of photosynthetic microorganisms than terrestrial C_3 and C_4 plants (Kaplan and Reinhold 1999) is primarily due to the action of extra- and intracellular carbonic anhydrase (CA), and the carbon concentrating mechanism (Van et al. 2001; Spalding et al. 2002; Vance and Spalding 2005). A number of CCM variants are now found among the different groups of photosynthetic microorganisms, evolved likely due to RuBisCO inefficiency (Douglas et al. 2003; Badger 1987) and its ability to bind O_2 as well as

© The Author(s) 2014
S.K. Singh et al., *Photosynthetic Microorganisms*, SpringerBriefs in Materials,
DOI 10.1007/978-3-319-09123-5_1

CO_2 at the same site. There have been significant major advances in our knowledge and understanding, at the molecular, mechanistic, and regulatory level of the photosynthetic microorganism's CCMs (Whitney et al. 2011). This process uses an active inorganic carbon (C_i; CO_2 and/or HCO_3^-) uptake system which leads to the internal accumulation of C_i to levels significantly higher than extracellular levels. The CCM also possesses two enzymes: carbonic anhydrase (converts accumulated HCO_3^- to CO_2) and RuBisCO (fix the CO_2 into reducing forms) (Falkowski and Raven 1997). The efficiency of the process is further improved many times by encapsulating the RuBisCO into specialized metabolic modules, e.g., carboxysome in cyanobacteria and pyrenoid in eukaryotic microalgae (Rae et al. 2013; Meyer and Griffiths 2013). Thus, carbon concentrating mechanism plays a vital role to enhance CO_2 fixation and cell growth, especially in the proper functioning of RuBisCO and assimilation of carbon via ubiquitous Calvin cycle. However, among the photosynthetic microorganism, *Chlmaydomonas reinhardtii* is the only proteome that studied even with the subcellular proteome level (Tirumani et al. 2014; Jungnick et al. 2014). The effect of different CO_2 concentration, roles of functional proteins, and mechanism involved in CCM of other ecologically important groups of photosynthetic microorganism are still to be investigated so far, to establish the mechanism(s) and regulation of CCMs.

Modulating the CCMs may be crucial in the energetic and nutritional budgets of a cell, and a multitude of environmental factors can exert regulatory effects on the expression of the CCM components. As for environmental modulation, much has been achieved at the level of phenomena, but deeper understanding requires more knowledge of mechanisms and of regulation at the cellular and molecular level. This is an important area in view of the increasing CO_2 concentration, temperature, and acidity of the surface ocean and of inland water bodies. In fact, world's rapid industrialization has led to severe global environmental challenges such as energy crisis, pollution, and global warming which are proportionally interlinked. It is widely recognized that fossil fuel combustion and other emissions resulting from anthropogenic interventions produce greenhouse gases in which carbon dioxide (CO_2) is the biggest contributor by volume (76 % of CO_2 emissions from Fuel Combustion, Olivier et al. 2013). It has been implicated in the global climate change and reducing them is a potential solution. In most cases, the easiest way to efficiently capture carbon is to catch it at high emission sources. Carbons capturing using photosynthetic microorganisms have been proposed as a potential solution to CO_2 capture transform into reducing carbon forms. In fact, photosynthetic microorganisms require CO_2 for their growth as one of the essential nutrients needed. Thus, they can consume tons of CO_2 for growth (Table 1.1) which results reduction in harmful greenhouse gases and helps in the global warming mitigation.

To date, many private companies such as Algaoil, Algenol, Algacake, Originoil, Sapphire, Solazyme Petroalge, and Originoil are directly involved in producing fuels from the algae. Algenol estimated that an acre of algae farm could utilize 60 tons CO_2 per year (Table 1.1). Whereas, one steps ahead, Algacake claimed a designed algae photobioreactor that can produce 10–12 times more algal biomass

Table 1.1 Production/
consumption of CO_2 per year

Produces/consumes	CO_2 (in tons per year)
Average person produces	2.3
Average car produces	6
An acre of normal forest consumes	2–3
An acre of oranges consumes	1–2
An acre of typical farm consumes	2
An acre of algae consumes (approx.)	60

(http://www.algenolbiofuels.com/Algenol%20101%20Sept%20
2009.pdf)

per unit area. Thus, 3-acre algae farm will consume up to 54,000 metric ton of carbon dioxide and produce 29 metric ton of biomass per year (http://www.algaecake.com/about.html). From these data, it seems that photosynthetic microorganism-based CO_2 mitigation could provide an excellent opportunity to sequester CO_2 from flue gas emissions from industrial sources and enhances atmospheric restoration by releasing oxygen that is chemically bonded in CO_2, back into the atmosphere. As an additional benefit, proposed scheme also recycles the CO_2 emissions into high-value commercial products that will offset the capital and the operation costs of the process. Increasing understanding of the mechanisms of CCMs, especially at the molecular level, will help our understanding of their mechanism of CO_2 fixation. Thus, the aim of this brief is to provide in-depth scientific details, in a comprehensive but concise way, on basic and applied aspects of carbon concentrating mechanism of photosynthetic microorganisms as well as emphasizing its integration with different industries. With the increasing demands for sustainable energy sources, photosynthetic microorganism with efficient CO_2 concentrating mechanisms is attractive models for biotechnological and transgenic applications for many existing industries such as biofuel, agriculture, wastewater treatment, food, and pharmaceuticals.

References

Badger MR (1987) The CO_2-concentrating mechanism in aquatic phototrophs. The biochemistry of plants: a comprehensive treatise (USA)

Brueggeman AJ, Gangadharaiah DS, Cserhati MF, Casero D, Weeks DP, Ladunga I (2012) Activation of the carbon concentrating mechanism by CO_2 deprivation coincides with massive transcriptional restructuring in *Chlamydomonas reinhardtii*. The Plant Cell Online 24(5):1860–1875

Douglas S, Larkum AWD, Raven JA (2003) Photosynthesis in algae. Springer, Netherlands

Falkowski PG, Raven JA (1997) Aquatic photosynthesis. Blackwell Science, Hoboken

Jungnick N, Ma Y, Mukherjee B, Cronan JC, Speed DJ, Laborde SM, Longstreth DJ, Moroney JV (2014) The carbon concentrating mechanism in *Chlamydomonas reinhardtii*: finding the missing pieces. Photosynth Res 121(2–3):159–173

Kaplan A, Reinhold L (1999) CO_2 concentrating mechanisms in photosynthetic microorganisms. Annu Rev Plant Biol 50(1):539–570

Lewy Z (2013) Life on earth originated where later microbial oxygenic photosynthesis precipitated banded iron formation, suppressing life diversification for 1.4 Ga. Int J Geosci 4(10):10

Meyer M, Griffiths H (2013) Origins and diversity of eukaryotic CO_2-concentrating mechanisms: lessons for the future. J Exp Bot 64(3):769–786

Olivier JGJ, Janssens-Maenhout G, Peters JAHW (2013) Trends in global CO_2 emissions: 2012 report. PBL Netherlands Environmental Assessment Agency

Rae BD, Long BM, Whitehead LF, Forster B, Badger MR, Price GD (2013) Cyanobacterial carboxysomes: microcompartments that facilitate CO_2 fixation. J Mol Microbiol Biotechnol 23(4–5):300–307

Spalding MH, Van K, Wang Y, Nakamura Y (2002) Acclimation of *Chlamydomonas* to changing carbon availability. Funct Plant Biol 29(3):221–230

Starr C, Taggart R, Evers C, Starr L (2012) Vol 2—Evolution of life. Cengage Learning, Boston

Tirumani S, Kokkanti M, Chaudhari V, Shukla M, Rao BJ (2014) Regulation of CCM genes in *Chlamydomonas reinhardtii* during conditions of light-dark cycles in synchronous cultures. Plant Mol Biol 85(3):277–286

Van K, Wang Y, Nakamura Y, Spalding MH (2001) Insertional mutants of *Chlamydomonas reinhardtii* that require elevated CO_2 for survival. Plant Physiol 127(2):607–614

Vance P, Spalding MH (2005) Growth, photosynthesis, and gene expression in *Chlamydomonas* over a range of CO_2 concentrations and CO_2/O_2 ratios: CO_2 regulates multiple acclimation states. Can J Bot 83(7):796–809

Whitney SM, Houtz RL, Alonso H (2011) Advancing our understanding and capacity to engineer nature's CO_2-sequestering enzyme. RuBisCO Plant physiol 155(1):27–35

Chapter 2
Carbon-Concentrating Mechanism

2.1 Introduction

None of the element on earth is more essential to life than carbon. Every living molecular machine is constructed across a middle staging of organic carbon. Unfortunately, carbon in the planet is locked in extremely oxidized structures, such as carbonate minerals (calcite, aragonite, etc.) and CO_2 gas (Walker 1985). In order to be functional, these oxidized structures ought to be unlocked and transformed into more organic forms, rich in C=C bonds and decorated with hydrogen atoms. With the help of light energy of sun, photosynthetic organisms perform this central task of carbon transformation in nature through the process called "photosynthesis." Among photosynthetic organisms, photosynthetic microorganisms (such as cyanobacteria and microalgae) play a significant role in the formation of organic biomass and oxygenic environment on Earth. They generate nearly half of the primary products of biosphere by contributing a large portion of carbon capture (Falkowski and Raven 1997). Majority of photosynthetic microorganisms undertake photosynthesis in an aquatic environment of ocean where they face a number of unique restraints regarding the efficient operation of carbon fixation through photosynthesis. In response to ancient changes in atmospheric CO_2 and O_2 levels, the photosynthetic microorganisms evolved a unique environmental adaptation, known as a CO_2-concentrating mechanism (CCM) (Badger and Price 2003), which has a significant positive effect on photosynthetic performance. In recent years, a deeper understanding of the mechanisms and genes underlying the operation of CCM has been increased rapidly.

2.2 Photosynthesis: Basis of Life on Planet

Photosynthesis (Greek: phos "light" and syntithenai "put together") is a central route in the global carbon cycle which serves as single prevalent flux of organic carbon in biosphere (Kirk 1994). It is a complex physico–chemical process occurs in a diverse group of organisms by which light energy from sunlight is absorbed

© The Author(s) 2014
S.K. Singh et al., *Photosynthetic Microorganisms*, SpringerBriefs in Materials,
DOI 10.1007/978-3-319-09123-5_2

by pigments and converted into chemical energy in the form of organic carbohydrates using carbon dioxide (CO_2) and water (See Eq. 2.1).

$$6CO_2 + 12H_2O + \text{Nutrients} + \text{Sun Light} \rightarrow C_6H_{12}O_6 + 6O_2 + 6H_2O \quad (2.1)$$

The photosynthesis process occurs widely in green pigments containing plants, algae, photosynthetic bacteria, and aerobic anoxygenic phototrophic bacteria which results in the release of molecular oxygen and the removal of CO_2 from the atmosphere that is used to synthesize carbohydrates (Shiba et al. 1979; Yurkov and Beatty 1998).

2.2.1 Basic Mechanism of Photosynthesis

The photosynthesis process encompasses two universal phases (Fig. 2.1).

In the first phase, "light-dependent reactions" involve light absorption, water splitting for electrons and protons source, generation of energy currencies such as NADPH and ATP and oxygen release as a by-product (Kirk 1994). The high-energy chemical intermediates, ATP and NADPH, further utilized as an energy source for the sequence of second phase "light-independent reactions" to fix CO_2 and reduce C_i in triose phosphates (carbohydrate precursors). Over all view of the whole photosynthesis process is shown in Fig. 2.2.

2.2.1.1 Light-Dependent Reactions

The light-dependent reactions are a sequence of chemical reactions occurs at the concentrated stacks of thylakoid called grana. It required the straight energy from

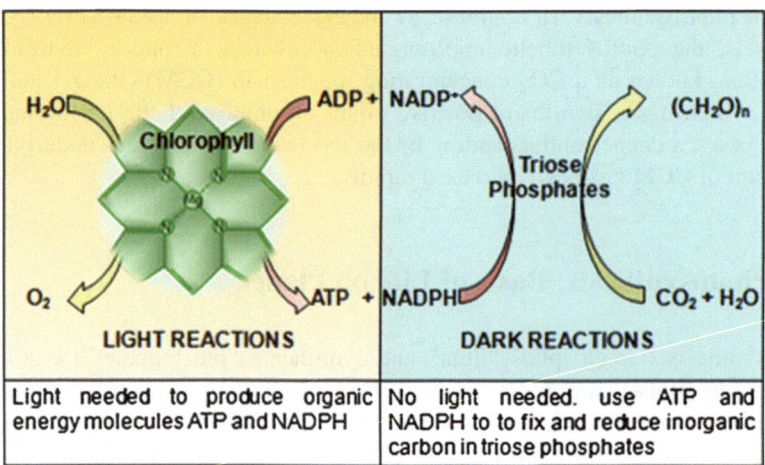

Fig. 2.1 Universal phases of photosynthesis process

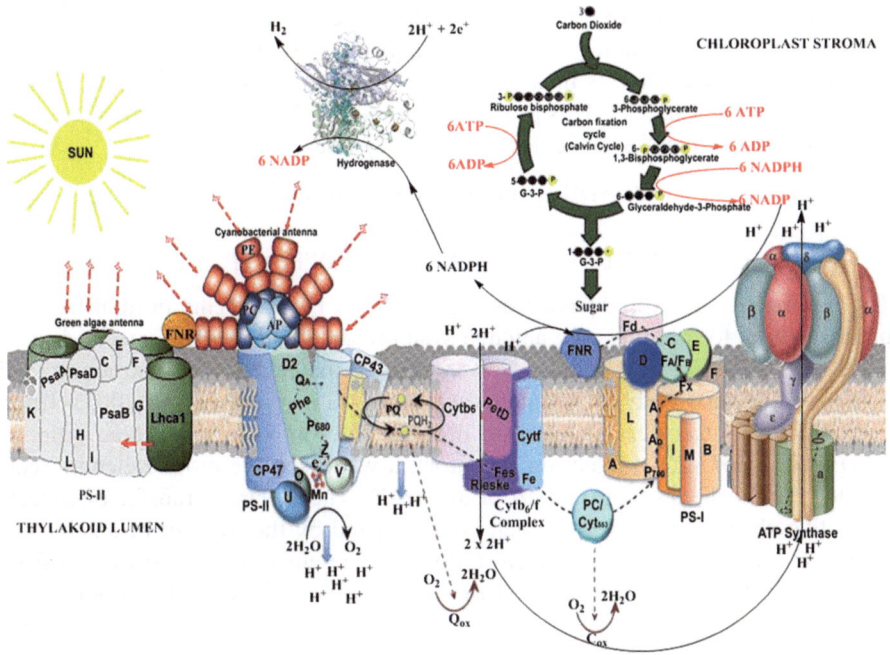

Fig. 2.2 Detailed view of light and dark reaction of photosynthesis

sunlight to make energy-carrier molecules (ATP and NADPH), which further used later in the light-independent reactions. The energy to drive light-dependent reactions comes from two photosystems: Photosystem II (PSII/P_{680}) and Photosystem I (PSI/P_{700}) (Barber 2003). The core of a photosystem is made of three molecules: a chlorophyll molecule, an electron acceptor (e.g., pheophytin), and an electron donor (e.g., water) (Van Gorkom 1985). All photosynthetic organisms have chlorophyll a and accessory pigments which include chlorophyll b (also c, d, and e in algae and protistans), xanthophylls, phycocyanin and phycoerythrin (cyanobacteria), and carotenoids (such as β-carotene) (Dufossae et al. 2005). Accessory pigments help to absorb light energy that chlorophyll a does not absorb. Initially, at PSII, light energy is absorbed by a chlorophyll molecule, an electron gains energy and is "excited" (photoexcitation). The excited electron is further transferred to a primary electron acceptor (Scholes and Fleming 2005), leaving a positively charged chlorophyll ion (photoionization). The positively charged chlorophyll ion then takes a pair of electrons from a splitting of two molecules of water (oxidized, i.e., loses electrons) into one molecule of molecular oxygen (see Eq. 2.2).

$$2H_2O \rightarrow 4H^+ + O_2 + 4e^- \text{(photolysis)} \tag{2.2}$$

The four electrons removed from the water molecules are transferred by an electron transport chain to ultimately reduce two molecule of nicotinamide adenine dinucleotide phosphate (NADP$^+$ to NADPH) (see Eq. 2.3). During the electron transport process, a proton gradient is generated across the thylakoid membrane. This proton motive force is then used to drive the synthesis of ATP (photophosphorylation) (see Eq. 2.4). This process requires PSI, PSII, cytochrome b$_6$f, ferredoxin-NADP$^+$ reductase, and chloroplast ATP synthase.

$$nADP + nP_i + photons \rightarrow nATP \qquad (2.3)$$

$$NADP^+ + 2e^- + 2H^+ \rightarrow NADPH + H^+ \qquad (2.4)$$

The final stage of the light reactions is catalyzed by PSI. This protein has two main subunits forming its core antenna system, *psa*A and *psa*B. A special pair of chlorophyll a molecules, denoted as P$_{700}$, lies at the center of the structure, and absorbs light maximally at 700 nm wavelength (Green and Parson 2003). Upon excitation, P$_{700}$ transfers an electron through chlorophyll and a bound quinone (Q$_A$) to ferredoxin (F$_d$) (electron acceptor), a water soluble mobile electron carrier located in the stroma (Gilbert et al. 2012). The electron transfer constructs a positive charge on the P$_{700}$, which is neutralized by the transfer of an electron from a reduced plastocyanin (electron donor). The electron transport chain from PSII to cytochrome b$_6$f to PSI is known as Z-scheme as redox diagram looks like letter Z.

2.2.1.2 Light-Independent Reactions

In the light-independent process (the Dark reaction), CO_2 from the atmosphere or water (for aquatic photosynthetic organisms) is captured and subsequently transformed by the addition of hydrogen to reduced carbon form such as carbohydrates (Schuster et al. 1984). The energy for this process comes from the first phase of the photosynthetic process. The incorporation of CO_2 into organic compounds is known as carbon fixation which we will discuss in more detail later in the chapter.

   Although photosynthesis process occurs in tiny micron-sized cells or organelles, it has a profound impact on the world's atmosphere and climate (Whitmarsh 1999). Each year, this CO_2 anabolic process helps in the transformation of approximately 100 billion tons of atmospheric carbon, which corresponds to almost 15 % of carbon in atmosphere (Raines 2011). Only within the last couple of decades, recent resurgence in basic and applied research on photosynthesis has been driven in part by recognition of novel strategies for compartmentalizing and enhancing the rates of photosynthetic carbon fixation reactions in a species-independent manner. Among all photosynthetic organisms, aquatic photosynthetic microorganisms face several challenges in acquiring CO_2 from the environment. Knowledge of CO_2 fixation in aquatic photosynthetic organisms is vital to understand the ecology between aquatic photosynthetic organisms and earth's atmosphere, for maintaining the equilibrium of organic carbon in biosphere.

2.3 Carbon-Concentrating Mechanism (CCM): A Potential Tool to Sequester Carbon

Carbon-concentrating mechanism (CCM) is a remarkable adaptation, evolved to maximize photosynthetic efficiency of many photosynthetic organisms in low-CO_2 (L-CO_2) environment. Since their role was first discovered (Badger et al. 1980), the mechanisms assisting the endurance of photosynthetic cells in L-CO_2 conditions have continued to be intensively studied. Concerns of sustainability, future food, and energy requirements are also motivating the researchers to elucidate the CCM machinery. Although in the last decade, significant progresses have been made to understand the exact CCM machinery of photosynthetic microorganism (Badger and Price 2003; Tabita et al. 2008; Whitney et al. 2011; Warlick 2013). However, many functional factors of CCMs are still unidentified or uncharacterized. The CCM is often an inducible process, and hence, the sensing of the lowering of CO_2 both intra- and extracellularly is needed to drive the structural and biochemical changes that accompany CCM induction. The signals that activate CCM have not been totally explored. More than one molecular inducer might exist to convey about the activation of genes and proteins central to the CCM process and the L-CO_2 adaptation of photosynthetic microorganism cells. The majority of evaluated aquatic CCMs are related to cyanobacteria and microalgae species (Raven et al. 2008; Reinfelder 2011, Ducat and Silver 2012; Barsanti and Gualtieri 2014).

2.3.1 Why Photosynthetic Microorganisms Need CCM?

Quite a number of photosynthetic microorganisms face the following key challenges of photosynthesis in aquatic environment.

2.3.1.1 Rate of Diffusion of CO_2

Since the energy transformations occurring in metabolism of living organisms are chiefly brought about by chemical changes in carbon-based biomolecules, the absorption and assimilation of CO_2 cannot be considered apart. In aqueous environment, the rate of diffusion of CO_2 is 10,000 times slower than the diffusion of CO_2 in air (Moroney and Ynalvez 2007). So that, there is relative equilibrium of CO_2 between air and water, which can result in carbon stress by causing a depletion of inorganic carbon (C_i) species including, CO_2, HCO_3^-, and CO_3^{2-} in water during active photosynthesis conditions. Furthermore, the photosynthetic microorganism growth environments are also subjected to fluctuations in C_i concentrations (CO_2 and HCO_3^-) due to pH (see Eq. 2.5)

$$CO_2 + H_2O \underset{\text{Slow}}{\overset{\text{pKa}_1=6.35}{\rightleftharpoons}} HCO_3^- + H^+ \underset{\text{Fast}}{\overset{\text{pKa}_2=6.35}{\rightleftharpoons}} CO_3^{2-} + H^+ \qquad (2.5)$$

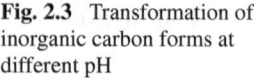

Fig. 2.3 Transformation of inorganic carbon forms at different pH

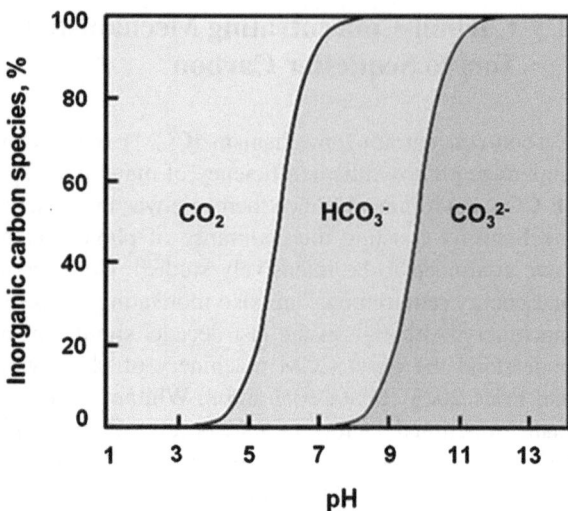

At the normal or slightly alkaline pH, water is typically low (approximately 10 μM) in CO_2 concentration (Lapointe et al. 2008), and causes the prevalence of less diffusible C_i form, HCO_3^-. It all affects the photosynthesis process that becomes carbon limited due to depletion of CO_2 from their instantaneous vicinity. The capability to scavenge CO_2 as rapidly as it becomes accessible is extremely advantageous to aquatic photosynthetic microorganisms. Figure 2.3 depicts aspects of this supply problem.

2.3.1.2 Limitations of RuBisCO

Despite its essential part in carbon fixation ability of photosynthetic organisms, ribulose bisphosphate carboxylase–oxygenase (RuBisCO) is, however, an unusual slow enzyme with a low affinity for CO_2. The catalytic ineffectiveness of RuBisCO originates not only from its low turnover rates but also exacerbated by O_2, being a competitive substrate of CO_2 in two competing reactions, carboxylation and oxygenation (Portis and Parry 2007). At atmospheric concentrations of CO_2, RuBisCO can only function at about one fourth of its catalytic capacity (Moroney and Ynalvez 2007). The existence of oxygen as a prevalent competitive substrate of CO_2 causes the redirection of fixed carbon into the photorespiratory cycle directing to the loss of at least 30 % of carbon fixed by RuBisCO (Raines 2011). As a result, all different CCMs form of aqueous photosynthetic organisms have evolved adaptations with a common aim of elevating CO_2 around RuBisCO to alter the CO_2/O_2 ratios at the active site in favor of the carboxylase reaction. In this manner, CCMs minimize the costly investment of metabolic energy and carbon in the photorespiration (Raven et al. 2008).

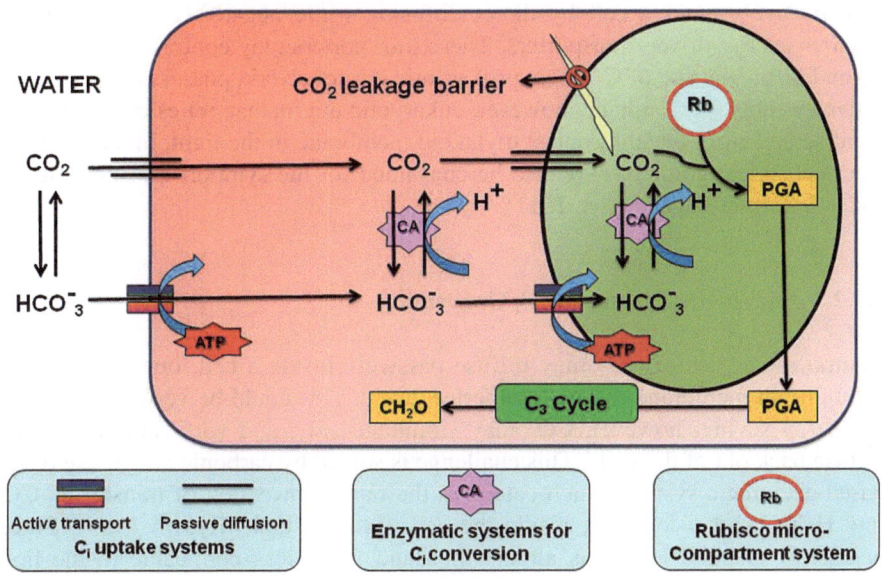

Fig. 2.4 Elements of photosynthetic microorganism CCMs

2.3.2 Functional Elements of Photosynthetic Microorganism CCMs

Even though, the complexity of cellular components and abilities of CCMs varies in different organisms, they have three major operational systems in common that allow them to achieve an effective use of CO_2. These are depicted in Fig. 2.4 and include the following.

2.3.2.1 Inorganic Carbon (C_i) Uptake Systems

The C_i uptake system*s* play a central role to achieve a satisfactory rate of CO_2 fixation into the cells of photosynthetic microorganisms under limiting carbon conditions. Photosynthetic microorganisms can use both form of C_i, ionic bicarbonate ions, and neutral CO_2 molecules. The maneuver of the C_i uptake systems enhances cytosolic concentrations of C_i to thousand times greater than extracellular concentrations (Daley et al. 2012). The C_i uptake system in photosynthetic microorganisms comprised of different HCO_3^- transporters and CO_2 uptake systems. Uptake machinery assists in the transfer, accumulation, and utilization of C_i as well as also prevents the diffusive leakage of CO_2 from actively photosynthesizing cells (Mukherjee 2013). The mechanism of C_i uptake is different in prokaryotes and eukaryotes photosynthetic microorganisms. Neutral molecules of CO_2

may passively enter a cell by direct diffusion while negatively charged HCO_3^- requires energy-driven transporters. These transporters may contrast in their affinity and diffusion rate of C_i. Most prokaryotic cyanobacteria possess energy-driven active systems for C_i uptake, however, eukaryotic microalgae relies on the pH gradient setup across the chloroplast thylakoid membrane in the light. In addition, the electrochemical gradient may also be consumed for the symport transportation in eukaryotic microalgae (Price 2011).

2.3.2.2 Enzymatic System for C_i Conversion

Neutral molecules of CO_2 may diffuse passively inside a cell, due to their high solubility in membrane lipids (Mukherjee 2013). This could be very nice in terms of energy saving; however, CO_2 may simultaneously, and with same efficiency, diffuse back out of the cells. This challenge is solved by carbonic anhydrase (CA)-based enzymatic system which catalyzes the rapid conversion of transferred CO_2 into HCO_3^- and vice versa, inside the cell (Tsuzuki and Miyachi 1989). Since, negatively charged HCO_3^- is almost thousand times less permeable to the lipid membranes than neutral CO_2 species (Price 2011). Thus, particularly at 7.8–8.2 cytoplasmic pH, CO_2 could be the ideal form of C_i for assimilation in cell while HCO_3^- form helps in preventing the outward leakage of C_i back into the medium. Various active CAs are localized at various sites in the cell predominantly close to the vicinity of RuBisCO where the level of CO_2 essentially to be elevated for proper CCM working (Mukherjee 2013).

Carbonic Anhydrases

The CAs (carbonate hydrolyase, E.C. 4.2.1.1) are ubiquitous metalloenzymes (mainly Zn) that catalyzes the quick reversible hydration reaction of CO_2 to HCO_3^- and protons (H^+) or vice versa. This "reverse" reaction gives CA its name, because it removes a water molecule from carbonic acid (see Eq. 2.6).

Carbon Di Oxide	Water	Carbonic Acid	Bicarbonate ion	Hydrogen ion

$$(2.6)$$

Due to the vital biocatalyst role of CAs, nature has advanced its catalytic ability as one of the fastest of all enzymes, to hydrate carbon dioxide and dehydrate bicarbonate a number of times (turnover number $\geq 10^4$–10^6 reactions per second)

(Donaldson and Quinn 1974). The equilibrium of both C_i forms in the solution obeys Henderson–Hasselbalch equation (see Eq. 2.7), and proportion of each form in solution is a function of pH.

$$pH = pKaH_2CO_3 + \log_{10}\left(\frac{[HCO_3^-]}{[H_2CO_3]}\right) \qquad (2.7)$$

The reaction shifted toward the formation of CO_2 in acidic conditions (pH < 6.4) while HCO_3^- form is prevalent in alkaline conditions (pH ~ 6.4 and 10.3).

(a) Basic structure and mechanism of action

The X-ray crystallographic data of CAs suggest that the zinc containing active site is situated at 15 Å deep cleft and coordinated with three histidines in a distorted tetrahedral geometry and one water/hydroxide molecule (Zn^{II}–H_2O or OH) (Zastrow and Pecoraro 2013). The binding cleft has hydrophobic and hydrophilic faces.

The Fig. 2.5 shows the structural arrangement of CA-II from protein data base (PDB) entry 1CA2 (Eriksson et al. 1988). The active site consists of a zinc prosthetic group, shown with a big red and green sphere. Imidazole rings of highly conserved histidine residues (shown with numbers 94, 96, and 119) directly coordinate with zinc while additional fourth coordination position, shown with small red sphere, occupied by water or hydroxide ion molecule (depending on medium pH).

At this specific pocket, atoms of threonine (Thr_{199}) and glutamate (Glu_{106}) as well as histidine assist to charge the zinc with a hydroxyl ion (Zastrow and Pecoraro 2013). The catalytic site also have affinity compartment for CO_2, bringing it close to the hydroxide group. CO_2 is not coordinated to the Zn^{II} but instead binds weakly ($K_d \approx 100$ mM) at a hydrophobic region (Krishnamurthy et al.

Fig. 2.5 Structural organization of carbonic anhydrase (PDB-1CA2)

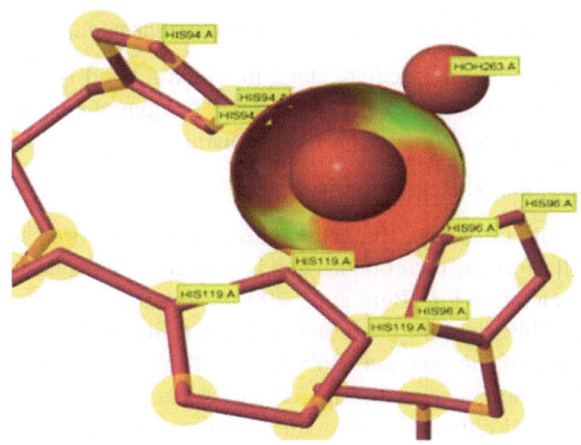

2008). Because zinc ion is a positively charged, it stabilizes the negative hydroxyl ion, thus it is prepared to attack the CO_2. Several CA isozymes have variations in these and other residues at active site, which may elucidate in their catalytic activity. Despite significant structural variations at the active sites, all CAs (majorly α- and γ-class) employ zinc hydroxide-binding mechanism (Parkin 2004). The zinc hydroxide-binding mechanism can be divided into four steps.

(i) Step 1: Deprotonation

As a universal feature of all known Zn^{II}-metalloenzymes, the Zn^{II} ion acts as a key element to activate this water molecule for catalysis. In biological machinery, zinc is always found only in the Zn^{+2} oxidation state (Kröncke and Klotz 2009). The chemical reaction of zinc elements linked with their positive charges and their capacity to form strong but kinetically labile bonds in more than one oxidation state (Berg et al. 2002). Core zinc metal helps in the activation of CA via release of a proton from a Zn-bound water (Zn^{II}–OH_2^+) to produce a zinc-bound hydroxide ion (Zn^{II}–OH). The role of the Zn^{+2} is here to lower the pK_a of the bound water molecule from 15.7 to 7, making the oxygen slightly more negative and polarization of hydrogen-oxygen bond. This is important for the mechanism, since the hydroxyl ion bound to Zn^{+2} form is more active than a water molecule.

(ii) Step 2: Carboxylation

Further, Zn^{II}–OH complex is also employed in H-bond interactions through the hydroxyl group of Thr_{199}, which is consecutively bonded with the carboxylate group of Glu_{106} (Kumar et al. 2007). These interactions make Zn^{II}–OH complex a potent catalytically active electron-rich nucleophile, and orient the substrate CO_2 molecule in a favorable position for the nucleophilic attack (Wang et al. 2011). This strong Zn^{II}–OH nucleophile attacks the CO_2 molecule bound in a hydrophobic pocket (substrate-binding site comprises residues Val_{121}, Val_{143}, Val_{207}, Leu_{198}, and Trp_{209} in hCAII) located above and to the right of the active site, leading to the formation of bicarbonate coordinated to Zn^{II}.

(iii) Step 3: Construction of ring-like intermediate complex

This is an intermediate stage in which a bond is created, connecting the hydroxyl ion and the CO_2 molecule. Anionic oxygen of CO_2 molecule forms a bond through core Zn^{II} to build a ring-like resonance.

(iv) Step 4: Regeneration of active form

In the final step, the active site of CA is restored for another round of catalysis. The formed bicarbonate ion is liberated, and one more water molecule binds to center zinc ion (Berg et al. 2002). Addition of water displaced the bicarbonate ion and liberated into solution. It leads to the formation of catalytically inactive acid form of the enzyme, Zn^{II}–OH_2^+. To regenerate the active form, a proton transfer reaction from the active site to the environment takes place, which may be assisted either by active site residues (such as His_{64}) or by buffers present in the medium.

The process may be schematically represented by reactions 1–4 in Fig. 2.6:

(b) Variants of CAs

Over the past decade, importance of CAs proteins and their genes in all domains of living organic world are implied by widespread distribution from prokaryotes such as archaebacteria and eubacteria to eukaryotes such as vertebrates (including humans) (Pastorek et al. 1994), invertebrates (Ferguson et al. 1937), and plants (Badger and Price 1994). The widespread variants of CA enzyme in nature is probably due to the fact that its substrate, C_i, is the most important element involved in all vital cellular processes. The comparison of amino acid sequences of all currently known CAs revealed the fact that they belong to five independent CA gene families, designated as α, β, γ, δ, and ε (Aspatwar et al. 2010). The crystal structures for representatives of α-, β-, and γ-classes have now been determined and shown in Fig. 2.7, while the structures of the recently identified δ- and ε-classes are yet to be solved.

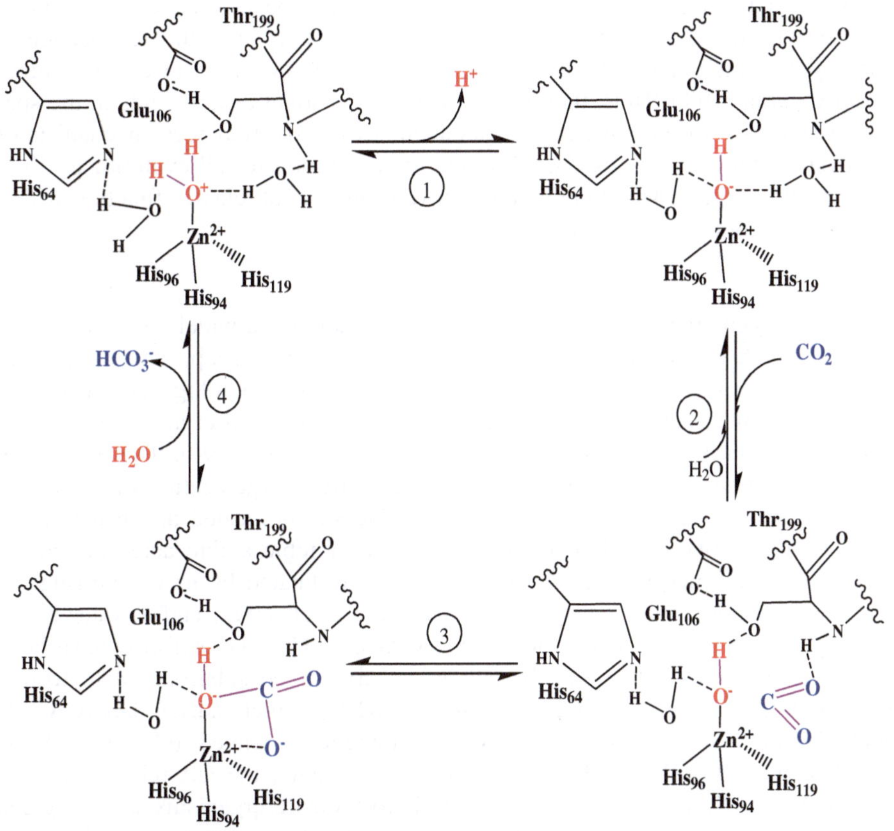

Fig. 2.6 Zinc hydroxide-binding mechanism of carbonic anhydrase

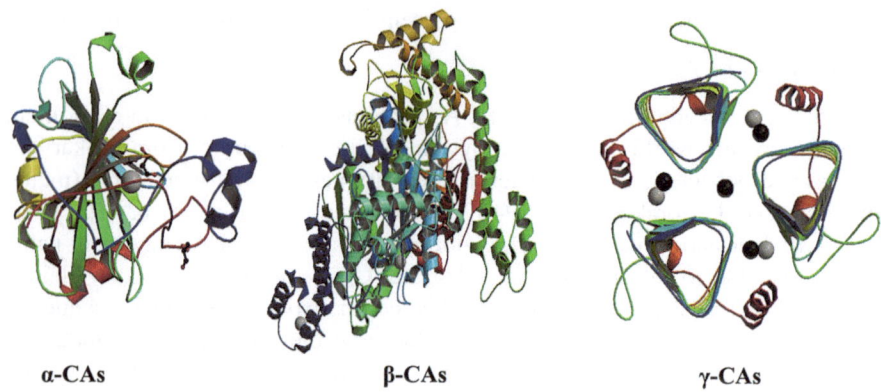

α-CAs   β-CAs   γ-CAs

Fig. 2.7 Refined structure of *α-CA* (PDB:1CA2, Eriksson et al. 1988), *β-CA* (PDB:2FGY, Heinhorst et al. 2006), and *γ-CA* (PDB:3KWD, Peña et al. 2010)

All CA enzyme families are of ancient origin and appears to have evolved independently from one another, i.e., no significant homology in amino acid sequence, thereby providing an excellent example of convergent evolution of catalytic function (Supuran 2011). Thus, they have structurally distinct overall folds in the spatial organization of proteins. Yet, despite their structural differences, mechanism of catalysis of the majority of CAs is similar and common for all evolutionary lineages of the enzyme. Metal-containing active centers of all classes function with a single zinc atom that is essential for catalysis.

(i) α-CAs

The α-CAs gene family is the most intensively studied and widely distributed CA family, which considered as the youngest phylogenetic group. The α-class are found predominantly in animals (Boone et al. 2013), but homologs have also been identified in the bacterium *Neisseria gonorrhoeae* (Hiltonen et al. 1998) and the green alga *Chlamydomonas reinhardtii* (Fukuzawa et al. 1990). The active site of α-class has high attraction for Zn, and geometrical organization of conserved histidine residues favors Zn binding. The active site is located at the bottom of a 15-Å-deep active site cleft dominated by hydrophobic amino acid side chains at the base of which is a Zn^{2+} ion invariably coordinated by imidazole rings of three his ligands, His_{94}, His_{96}, and His_{119} (Tetu et al. 2007) and a water molecule/hydroxide ion. To date at least 15 α-CA- or α-CA-like isoforms have been found in mammals, which can be subdivided five broad subgroups as cytosolic CAs (CA-I, CA-II, CA-III, CA-VII, and CA XIII), mitochondrial CAs (CA-VA, and CA-VB), secreted CAs (CA-VI), membrane-associated (CA-IV, CA-IX, CA-XII and CA-XIV), and those without CA activity, the CA-related proteins (CA-RP VIII, X, and XI). In the aquatic photosynthetic organisms such as green alga *Chlamydomonas reinhardtii*, three CA isozymes located at periplasmic glycoproteins have been sequenced, evolutionary related to mammalian CAs (Moroney et al. 2011). α-CAs typically activated as protein monomers of about 30 kD that are mostly composed of 10-stranded secondary structure, a twisted β-sheet, which separates the α-CA molecules into two halves. Except for two pairs of parallel strands, the β sheet is antiparallel (Liljas et al. 1972).

(ii) β-CAs

The β-CAs are perhaps the most diverse lot of the five currently known CAs, structurally and functionally. The β-CAs were initially identified in chloroplast of higher plants (Burnell et al. 1990; Fawcett et al. 1990) but are now known to be present in various subcellular compartments of microalgae (Eriksson et al. 1996), cyanobacteria (Fukuzawa et al. 1992; Soltes-Rak et al. 1997), eubacteria (Hewett-Emmett and Tashian 1996), archaea (Smith and Ferry 1999), and fungi (Schlicker et al. 2009). Interestingly, β-CA is far more diverse in amino acid sequence than the other two classes, suggesting that they evolved independently. The entire β-CAs share an exclusive α/β fold not exist in any other proteins (Covarrubias et al. 2006). In contrast to α-CAs that is mostly composed of β-sheets, β-CAs contains a number of α-helices. Moreover, β-CAs have only one conserved histidine residue (His $_{205/459}$), whereas α-CAs have three (Supuran and Scozzafava 2007). Unlike α and γ-CAs which form strictly monomers and trimers, β-CAs are only functional when homodimeric active core of β-CA oligomerized as a dimer, tetramer, or octamer depending on the species of origin (Mitsuhashi et al. 2000; Huang et al. 2011). The oligomerization state appears to be motivated by outside extensions or exclusive amplifications of the secondary structure of basic β-CA fold (Kimber and Pai 2000; Krissinel and Henrick 2007). β-CA can adopt a variety of functional oligomeric states with molecular masses ranging from 45 to 200 kD (Mitsuhashi et al. 2000). However, the fundamental active structural unit of β-CAs appears to be a dimer or multimers. Dimerization enables formation of the hydrophobic pocket required for CO_2binding and forms the active site at the interface. Core zinc ion geometry is supported by a combination of cysteine, histidine, and glutamic acid or cysteine again, depending on the species (Mitsuhashi et al. 2000).

The most common arrangement for β-CA is a tetramer (HICA, ECCA, Rv1284) or a pseudo-tetramer composed of two pseudo-dimers (PPCA, HTCA) (Kanth et al. 2012). The extant X-ray crystal structures of β-CA appear to fall into two distinct structural classes as determined by the organization of the active site region in its uncomplexed state, designated here as "Type I β-CA" and "Type II β-CA" (Sawaya et al. 2006). The principal differences between these two types of β-CA relate to the ligation state of the active site zinc ion, and the orientation and organization of nearby residues (Table 2.1).

(iii) γ-CAs

The γ-class may be the most prehistoric form of CAs, having evolved long before the α-class, to which it is more closely related than to the β-class (Badger and Price 2003). The γ-class is also broadly distributed in diverse species from all three domains of life, predominantly in bacteria and archaea domains. A γ-CA was first discovered and isolated in the methanogenic archaebacterium that grow

Table 2.1 Physiological characteristics of Type I β-CA and Type II β-CA

Physiological character	Type I β-CA	Type II β-CA
Coordination sphere for zinc	$Cys_2His(X)$, where X is an exchangeable ligand (e.g., acetate, acetic acid, water)	$Cys_2HisAsp$
Asp–Arg dyad (helps in orient the Asp residue to accept a hydrogen bond from an exchangeable ligand atom bound directly to the Zn^{2+})	Asp–Arg dyad present	A broken Asp–Arg dyad is present
Hydrogen bond donor (for interaction between Zn^{2+} and HCO_3^- ions in the exchangeable ligand position	Hydrogen bond donor present	Hydrogen bond donor absent
Narrow hydrophobic active site cleft that lies along the dimer or pseudo-dimer interface and leads to the active site Zn^{2+} ion	Present	Present

in hot springs *Methanosarcina thermophila* (Alber and Ferry 1994). To date, the only "Cam" (for **CA** of *Methanosarcina thermophila*) has been shown to have CA activity in trimerized form (Kisker et al. 1996). All other Cam homologs from both plants and bacteria, including CcmM from the cyanobacteria *Synechocystis* PCC6803 and *Synechococcus* PCC7942, were found to lack CA activity (Peña et al. 2010). This suggested that these homologs have evolved a different function and that Cam is a relic. They obtain energy for growth by metabolizing acetate to CH_4 and CO_2 (Smith and Mah 1978). The role of γ-CA in acetate metabolism is to drive forward reaction of acetate to methane and CO_2 by removing the CO_2 concentration by converting it to HCO_3^- outside the cell. γ-CAs catalyze the reversible hydration of CO_2 to HCO_3^- ion on opposing sides of the membrane thereby facilitating anion exchange, in order to reduce the concentration of CO_2 produced in acetate metabolism. In the cyanobacterium *Synechocystis*, the bifunctional CcmM protein localized in carboxysome shows an N-terminal γ-CA-like domain, which has been proposed to bind HCO_3^-/CO_2 (Cot et al. 2008). Recent work has indicated that γ- or γ-like CAs are part of Complex I of the mitochondrial electron transport chain in plants and algae (Wang et al. 2012).

According to the structural classification of proteins (SCOP), γ-CA is part of the Trimeric LpxA Enzyme superfamily, which is characterized by single-stranded polypeptides with left-handed beta-helix fold. Cross-sectional profiles of the γ-CA trimer reveal that each left-handed beta-helix monomer structure resembles an equilateral triangle complex (Iverson et al. 2000). The beta-helix consists of three untwisted, parallel sheets that are connected by left-handed crossovers. The active sites are located at the interfaces between two β-helices. The interface is stabilized by H-bonds, salt bridges, and hydrophobic interactions. The trimer contains 3 active sties, and each monomer contributes His residues located on the surface to coordinate with the 3 zinc ion, Zn^{2+} or the cobalt ion, Co^{2+} (Kisker et al.

1996), and one at each subunit interface. The unique feature of γ-CA is its ability to utilize both metal ions equivalently for its active site, depending on their availability. There are no significant differences between both forms (zinc-bound and cobalt-bound γ-CA) structures and their catalytic mechanism for carbon dioxide hydration reaction. Both have three histidine residues (His_{81}, His_{117}, and His_{122} residues) that coordinate the ion with the active site (Tripp et al. 2001). In addition to His, there are several other residues which have been identified to play an important role in the active site catalytic mechanism. For example, glutamine (Gln_{75}) and asparagine residues (Asn_{73}) have been found to help orient Co^{2+} ion for attack on the CO_2, and asparagine residue (Asn_{202}) prepares the CO_2, by polarization, for attack by the Co^{2+} ion. The reaction mechanism of the γ-class is similar to that of the α-class, even though, overall folds and active site residues are different (apart from those that ligand the zinc).

(iv) δ-CAs

A fourth class of CA named δ-class has been isolated from the marine diatom *Thalassiosira weissflogii* (Roberts et al. 1997). X-ray absorption spectroscopy of the δ-CA, *T. weissflogii* CA1 (TWCA1), has shown that it indeed does contain a Zn^{2+} ion bound by histidine residues. Presently, there are only 4 other proteins that display amino acid sequence similarity to TWCA1, and thus, its distribution may be restricted to only a small number of diatom species (So and Espie 2005).

(v) ε-CAs

The fifth epsilon class of CAs occurs exclusively in bacteria containing α-type carboxysomes in a few chemolithotrophic bacterium *Halothiobacillus neapolitanus,* hydrogen bacteria and many strains of marine cyanobacteria that contain CsoS3-carboxysomes (So et al. 2004; So and Espie 2005). X-ray 3-D crystal structure analyses of *H. neapolitanus* suggest that active site of ε-CA bears some structural resemblance to β-CA (Sawaya et al. 2006) particularly near the metal ion site with a histidine and two cysteine residues acting as zinc ligands, in spite of the absence of any primary sequence similarity. This suggested that CsoS3 is a subclass of β-CA comes from the striking structural similarity of the Zn^{2+}-containing active site and from the fact that both need to form dimers in order to be active (Sawaya et al. 2006). In all examples to date, CsoS3 is encoded within the Cso operon which encodes all the components for the α-carboxysome (Rae et al. 2013). Thus, the two forms may be distantly related, even though the underlying amino acid sequence has since diverged considerably. This class of CA has not been found in eukaryotes.

2.3.2.3   RuBisCO Micro-Compartment System

A third basic requirement of a CCM is the existence of effective RuBisCO-rich micro-compartment system, essential for fixing and minimizing the leakage of CO_2 (Price et al. 2008). RuBisCO is commonly, catalyzes the first rate limiting step within the Calvin cycle (Berg et al. 2002). RuBisCO has a large active site that can accept both CO_2 and O_2 as a substrate (van Lun et al. 2014). O_2 has a higher affinity for the active site and therefore will bind at a higher rate than CO_2 , decreasing the carbon fixation ability of RuBisCO. Photosynthetic microorganisms have developed small vesicles within the cell that have tightly packed and concentrated levels of RuBisCO molecules within proteinaceous shell of a specific structure known as the carboxysome (cyanobacteria) or pyrenoid (eukaryotic microalgae) (Mukherjee 2013). The CO_2 is shuttled from outside environment into the proteinaceous shell to maximize the concentration exposure to RuBisCO. We will discuss details of these proteinaceous shells in consequential chapters of the brief.

RuBisCO: Nature's CO_2-Sequestering Enzyme

RuBisCO (Ribulose-1,5-bisphosphate carboxylase/oxygenase, E.C.4.1.1.39), the most abundant protein on Earth (Ellis 1979), forms a bridge in between living biological system and the lifeless chemical network via converting inorganic of the air to organic carbon. It is believed that nearly all of the carbon atoms that are present in living organisms have passed through the active site of RuBisCO, as 95 % of all carbon fixations by C_3 organisms (that includes all phytoplankton) occur via RuBisCO (Raven 1997). RuBisCO occurs universally in most autotrophic organisms from prokaryotes (photosynthetic and chemoautotrophic bacteria, cyanobacteria and archaea) to eukaryotes (various algae and higher plants) on land and in the ocean (Andersson 2008). Since, it constitutes up to 50 % of the soluble protein in the leaf of C_3 plants and ~30 % in C_4 plants (Spreitzer and Salvucci 2002; Sugiyama et al. 1984), considered an extremely important enzyme ecologically, agriculturally, and industrially. Due to the importance and abundance of RuBisCO, aspects of the genetics, microbiology, molecular biology, biochemistry, and evolution of the enzyme have been studied intensely.

Molecular Forms of RuBisCO

The first RuBisCO structure to be solved was that from the bacterium *Rhodospirillum rubrum* (Andersson et al. 1989; Schneider et al. 1990). Further studies by many researchers (Portis 1992; Newman and Gutteridge 1993; Taylor and Andersson 1997; Sugawara et al. 1999; Tabita 1999; Portis 2003; Warlick 2013) have shown that all RuBisCOs found in nature are comprised of two types of polypeptide subunits: a large subunit (L) of 50–55 kD and a small subunit (S) of 12–16 kD (Andersson 2008). On the basis of their number, presence or absence, and structural arrangement, RuBisCOs can be classified into four different molecular forms designated as form I, II, III, and IV (Tabita et al. 2007) (see Fig. 2.8). These multimeric forms have often distinctive features; however the primary structural motif, frequent to every holoenzyme forms, is the catalytic large

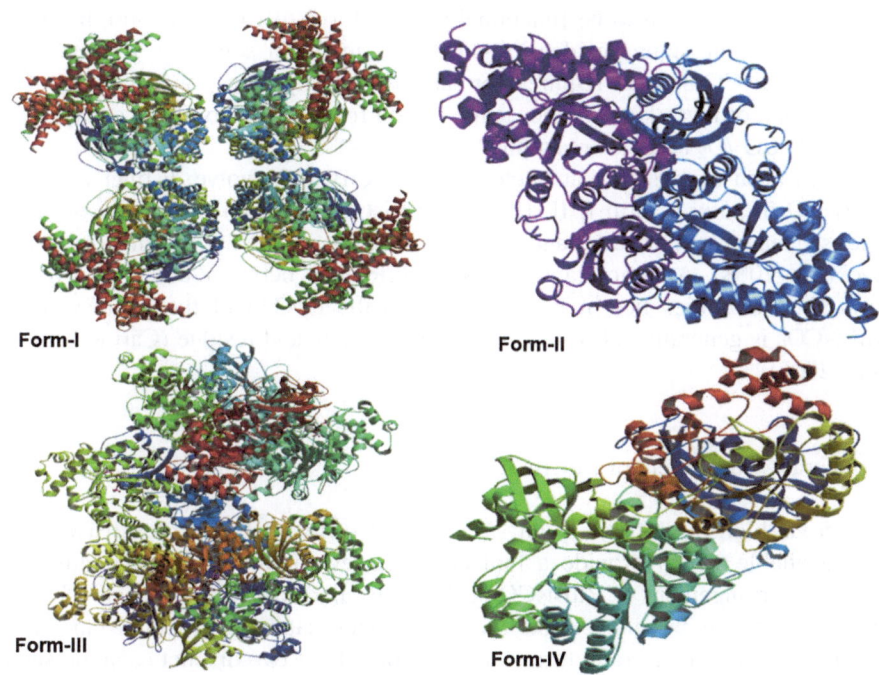

Fig. 2.8 Structure of RuBisCO forms; *form I* (Bracher et al. 2011), *form II* (Tabita et al. 2008), *form III* (Nishitani et al. 2010), and *form IV* (Tabita et al. 2007)

subunit dimer (Tabita et al. 2008). The most abundant form of RuBisCO, Form I is comprised of 8 small and 8 large subunits. The fundamental catalytic structural unit, the dimer of L, is polymerized 4 times to form a core (L_8) of 8 L subunits, with small subunits on top and bottom of this core (Saschenbrecker 2007; Saschenbrecker et al. 2007; Tabita et al. 2008). It is found in plants, algae and cyanobacteria and some members of α-, β- and γ-proteobacteria. Further categorization of form I of RuBisCO contains IA and IB forms (green-type enzymes from cyanobacteria, eukaryotic algae, and higher plants) and IC and ID forms (red-type enzymes from non-green algae and phototropic bacteria) (Tabita et al. 2008). Form II is composed of only the large subunit and is present in dinoflagellates and some members of α-, β- and γ-proteobacteria. Interestingly, the phototrophic purple non-sulfur bacteria (e.g., *Rhodobacter sphaeroides*, *Rhodobacter capsulatus*) and other organisms including *Hydrogenovibrio marinus* and some *Thiobacillus* species contain both form I and form II RuBisCO, and in *R. capsulatus* both forms are expressed under photoautotrophic conditions (Badger and Bek 2008). Form III is found in archaea and consists of a large subunit in a dimeric or pentameric arrangement (Andersson 2008). Form IV is referred to as the RuBisCO-like protein (RLP) because it does not catalyze bonafide RuBisCO CO_2/O_2 fixation reactions by using RuBP as the substrate (Hanson and Tabita 2001). Even though the RLPs do not seem to be functionally related to RuBisCO, they do, however, share a common sequence identity but do not maintain some key residues present in RuBisCO that are required for catalytic activity. Thus, one may speculate that the lack of these key residues is the reason for RLP's inability to catalyze the RuBisCO CO_2/O_2 bonafide reaction.

It has been suggested that photosynthetic RuBisCO evolve distinct forms of RuBisCO in nature; form I, II, III, and RuBisCO-like form IV based on amino acid sequences, phylogeny, and structure (Tabita et al. 2007; Andersson and Backlund 2008). An important structural difference between the form I/II and form IV subfamilies lies in loop 6 in which the Lys334 of the photosynthetic RuBisCOs is generally substituted by another amino acid residue (Carrae-Mlouka et al. 2006).

Structural Arrangement of RuBisCO

Numerous high-resolution crystal structures of different forms of RuBisCO are now available which provide a molecular framework for the understanding of structural arrangement of RuBisCO at the molecular level. On the basis of available facts, it is now believed that the fundamental catalytic structural unit of all RuBisCOs is common to all forms, usually consists of two distinct catalytic subunits; large catalytic subunits (L, about 55 kD each) and small subunits (S, about 15 kD each) (Andersson 2008). In form I RuBisCO, eight copies each of two distinct subunits are cemented to form quaternary structure of about 560 kD molecular mass for the complete protein (L_8S_8) superstructure (Bracher et al. 2011). The 8L subunits of form I are arranged as an octameric core surrounded by two layers of four S subunits, with each layer located on opposite sides of the molecule. A Mg^{2+} cofactor as well as the carbamylation of Lys_{201} is also required for the activity of the enzyme (Tabita et al. 2008).

Genes for RuBisCO

The clustering of genes (e.g., *rbcL, rbcX,* and *rbcS*) is thought to assist in coding for the structurally related various complex components synthesis and assembly (Tabita 1999) of CO_2-fixing hexadecameric (L_8S_8) RuBisCO enzyme. For synthesis of the functional RuBisCO holoenzyme in microalgae, eight identical large subunits encoded by the chloroplast gene *rbcL* and eight identical, small subunits encoded by the nuclear gene *rbcS* (Clegg et al. 1997). Within cyanobacteria, both genes are adjacent and co transcribed.

(a) *rbcL*

Despite variations in the amino acid sequences (average amino acid sequence identity is 31 %), the overall secondary structural motifs of the large (catalytic) subunit shares similarities and well conserved within all forms of RuBisCO super family (Tabita et al. 2007). *rbcL* expression may be regulated by the epistasy of synthesis (CES) paradigm, in which unassembled L-subunit motifs connect to mRNA of *rbcL* to autoregulate its translation (Whitney et al. 2011). Large subunits within RuBisCO are arranged as antiparallel dimers, with the smaller N-terminal domain (4–5 stranded mixed β sheet) of one monomer adjacent to the C-terminal domain (8 consecutive β–α units) of the other monomer (Tabita et al. 2007). Each active site is at an interface between monomers within an L_2 dimer, explaining the minimal requirement for a dimeric structure. The β–α units of C-terminal, are linked by many loops of different length and arranged as an eight stranded parallel α/β barrel structure, which act as a evolutionary markers and directed evolution techniques to engineer novel catalytic activities (Vega et al. 2003). The substrate-binding site is at the intra-dimer ($RbcL_2$) interface on the mouth of a α/β-barrel domain of the large subunit, linking the C-terminal domain of the one large subunit (β-strands) and the N-terminal domain of the second large subunit. Consequently, the functional unit configuration of RuBisCO is an L_2 dimer of large subunits harboring two active sites. The substrate binds in an extended conformation across the opening of the α/β barrel and is secured at two distinctive phosphate-binding sites at reverse sides of the α/β-barrel and in the center at the Mg^{2+} cofactor-binding site (Tabita et al. 2007). Most catalytic residues at enzyme active site are polar, including some charged amino acids (e.g., Thr, Asn, Glu, and Lys) (Bartlett et al. 2002).

(b) *rbcS*

Small subunits are not necessary for the assembly of the $RbcL_8$ core. However, availability of *rbcS* (13.3 kD) protein up-regulates the gene expression of *rbcL* primarily at the transcript level in a quantitative manner for stoichiometric assembly of RuBisCO holoenzyme (Morita et al. 2014). It is tempting to speculate that the small subunits contribute substantially to the differences in kinetic properties observed among different RuBisCO enzymes. In eukaryotic microalgae, small subunits are probably encoded by the *rbcS* multigene family in the nuclear genome (Clegg et al. 1997). It is believed that mRNA of *rbcS* has been laterally transferred from the ancestral plastid's genome to become a nuclear multigene family (Whitney and Andrews 2001). They have a transit peptide, bearing an amino-terminal targeting signal, which helps small subunit precursor proteins to be imported and assembled into the chloroplasts after translation on cytosolic ribosomes, where they are processed and folded to the native state (Flores-Paerez and Jarvis 2013). Whereas, the large subunits display relatively small variations in the different forms, the small subunit is more diverse. The small subunits help to sustain the catalytic competence and structural integrity of $RbcL_8S_8$ holoenzyme by establishing prevailing links among the four $RbcL_2$ dimers (Windhof 2011, Liu et al. 2010). The common core structure of small subunits consists of a four-stranded antiparallel β-sheet covered on one side by two helices. Among the form I RuBisCOs, most striking variations occur in two distinct locations, the small subunits differ in between β strands A and B of small subunit, called βA–βB loop (Andersson 2008).

(c) *rbcX*

In contrast to eukaryotes, most prokaryotes such as cyanobacteria, the gene *rbcX* (~15.5 kD) is present between *rbcL* and *rbcS* and co-transcribed with the *rbcL* (52 kD) and *rbcS* genes on the same operon (Larimer and Soper 1993). Previous co-expression studies showed that *rbcX* product of the intermediary *rbcX* gene is not part of the final RuBisCO complex and unlike *rbcL* and *rbcS* whose sequences are highly conserved by functional constraints. The *rbcX* sequence is highly vari-able (<60 % similarity) among cyanobacterial species (Rudi et al. 1998). However, recent evidence suggests that juxtaposition of *rbcX* within an *rbcLXS* operon is highly conserved in β-cyanobacteria, recommending that the *rbcX* product may function in a role associated with CO_2 fixation (Emlyn-Jones et al. 2006; Saschenbrecker 2007; Saschenbrecker et al. 2007; Onizuka et al. 2004).

Reaction Mechanism of RuBisCO

RuBisCO is most commonly known to a bifunctional enzyme that occurs in the stroma of chloroplasts and catalyzes both carboxylation and oxygenation reactions (Tabita et al. 2008). These reactions are the basis for the name RuBP carboxylase/ oxygenase (RuBisCO). In order for RuBisCO to function, it requires substrates supplied from the surrounding environment. In these reactions, both CO_2 and O_2 substrates compete for the same active site on RuBisCO to drive photosynthesis and photorespiration, respectively. Both metabolic routes are found in most auto-trophic organisms, ranging from prokaryotes (cyanobacteria and other phototro-phic and chemoautotrophic bacteria) to eukaryotes (various algae and higher plants) (Spreitzer et al. 2002).

(a) *Carboxylation*

When CO_2 is the substrate gathered in specific location of the RuBisCO, it per-forms a conformation change and catalyzes the carboxylation reaction with the activation of RuBP carboxylase. Carboxylation involves the fixation of one mol-ecule of CO_2 with a molecule of five-carbon sugar substrate, ribulose-1,5-bis-phosphate (RuBP) to produce a highly unstable six-carbon reaction intermediate (Blankenship 2014). Due to the intermediate molecules instability, carbon splits into two molecules of 3-phosphoglycerate (3PGA). This reaction occurs in several partial reactions (Karkehabadi 2005) (see Fig. 2.9).

- Enolization—Enzymatic abstraction of a proton (H^+) from C-3 of the RuBP substrate results in the formation of the 2,3-enediol intermediate (I). Mg^{+2} aids in stabilizing the 2,3-enediol transition state for CO_2 addition and facilitates the C–C bond cleavage that leads to two 3-C products.
- Carboxylation—The addition of CO_2 at C-2 to create a 6-carbon β-keto acid intermediate (II), 2-carboxy-3-keto-arbinitol-1,5-bisphosphate (CKABP).

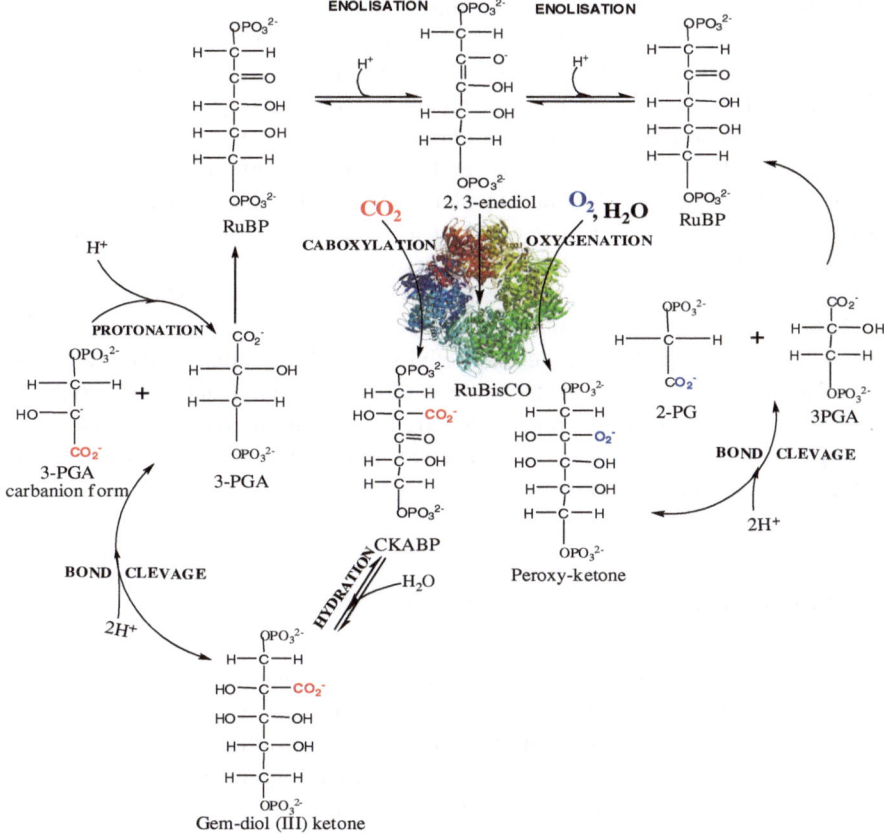

Fig. 2.9 RuBisCO pathways for carboxylation and oxygenation

- Hydration—The hydration of CKABP yields the gem-diol (III) form of the ketone.
- Deprotonation—Deprotonation at C-3 hydroxyl of the gem-diol (III) leads to C–C bond cleavage and results in formation of one molecule of 3-phosphoglycerates (3PGA) and one molecule of 3-PGA in the form of carbanion.
- Protonation—The carbanion is protonated, and the second molecule of 3-PGA is formed.

(b) *Oxygenation*

When molecular O_2 is the substrate, RuBisCO catalyzes the oxygenation of sugar substrate RuBP to yields one molecule each of 3PGA and 2-phosphoglycolate (Karkehabadi 2005). The phosphoglycolate has very limited use in most organisms and needs to be re-circulated through the sequence of complex energy-requiring reactions called C-2 photosynthesis or photorespiration that partly salvages carbon from 2-phosphoglycolate, via conversion to 3-phosphoglycerate, involves enzymes of chloroplasts, peroxisomes, and mitochondria. This pathway recovers 3/4 of the carbon from 2-phosphoglycolate as 3-phosphoglycerate while the rest is released as CO_2. As photorespiration consumes ATP and reducing power, while losing CO_2, before it is converted to PGA and reenters to the metabolic pathways for carbon fixation (Laing et al. 1974). It is consider a wasteful process which substantially reduces the efficiency of CO_2 fixation by up to 50 % (Ogren 1984). Thus, the incapability of the RuBisCO to avoid the reaction with O_2 greatly reduces the photosynthetic capacity of photosynthetic organism. It would appear that eliminating or reducing the RuBisCO oxygenase activity would potentially increase carbon assimilation, thereby rising photosynthetic efficiency significantly (McGrath and Long 2014). Many algae and photosynthetic bacteria have conquered this restriction by devising means to raise the CO_2 concentration around the enzyme through carbon-concentrating mechanism.

2.4 Fate of Carbon in Photosynthetic Microorganisms

Although CO_2 occurs in small amounts in the atmosphere, it has a considerable impact on living organisms, since it is a key substrate of photosynthesis. The aquatic environment is home of diversity of photosynthetic pathways as terrestrial environments, and therefore, photosynthetic organisms reduce CO_2 via various carbon fixation mechanisms, C_3, C_4, CAM, and C_3–C_4 photosynthetic pathways (Xu et al. 2012). Reduction takes place in in the stroma, or soluble phase, of chloroplasts, coupled to the consumption of NADPH and ATP synthesized by the light reactions of thylakoid membranes (Blankenship 2014). Here, CO_2 and water are combined with ribulose-1,5-bisphosphate to form two molecules of 3-phosphoglycerate. Phosphoglycerates are familiar molecules in the cell, and many pathways are available to use it to produce larger biomolecules such as carbohydrate. Most of the phosphoglycerate made by RuBisCO is recycled to build more ribulose bisphosphate, which is needed to feed the carbon-fixing cycle. The continued operation of these cycles is ensured by the regeneration of ribulose-1,5-bisphosphate. From many studies on primary photosynthetic carbon metabolism, it is believed that the operation of the Calvin–Benson cycle (C_3 cycle) is predominant in algae and cyanobacteria. However, recent papers have also reported evidence for the operation of C_4 photosynthesis and both C_3 and C_4 fixation in some species, as an alternative CCM. Alterations of photosynthetic pathways under environmental stress such as CO_2 deficiency have been suggested to contribute to the adaptation of photosynthetic organisms to environmental stress. The major physiological differences between C_3 and C_4 cycles are tabulated in Table 2.2.

Table 2.2 Physiological differences between C_3 and C_4 cycles

Property	C_3 cycle	C_4 cycle
CO_2 molecule acceptor	Ribulose bisphosphate	Phosphoenol pyruvate
First stable product	Three-carbon compound 3-phosphoglycerate (3PGA)	Four-carbon compound 3-oxaloacetic acid (OAA)
Photorespiration rate	High and leads to loss of fixed CO_2	Negligible or almost absent
Optimum temperature	20–25 °C	30–45 °C

2.4.1 C₃ or Calvin–Benson–Bassham Cycle

A majority of photosynthetic organisms assimilate CO_2 via a set of redox reactions, C_3 pathway (Calvin–Benson–Bassham cycle) that occurs without light during photosynthesis (Björn 2008). The cycle was elucidated about 50 years ago as a result of a series of elegant experiments by Calvin, Bassham, and Benson at the University of California, Berkeley, for which a Nobel Prize was awarded in 1961 (Calvin et al. 1950). They used radioactive $^{12}CO_2$ isotopes to reveal the path of carbon atoms taking place in unicellular green alga *Chlorella pyrenoidosa,* during the transformation of CO_2 into carbohydrates. The C_3 cycle utilizes the high-energy products of light-dependent reactions, ATP and NADPH, to fix atmospheric CO_2 into carbon compounds that are used to fuel the rest of plant metabolism (Whitmarsh 1999). The carbon in CO_2 is the most oxidized form, (+4) oxidation state, found in nature. With 4 valence shell electrons, carbon tends to form covalent compounds. Thus, first stable intermediate, 3-phosphoglycerate of the C_3 cycle, is more reduced (+3) and after it is further reduced to glyceraldehyde-3-phosphate (+1) product (Blankenship 2014). Therefore, early reactions of the C_3 cycle complete the reduction of atmospheric carbon and, in so doing, facilitate its incorporation into organic compounds. This cycle operates in plants, algae, cyanobacteria, some aerobic or facultative anaerobic proteobacteria, CO_2-oxidizing mycobacteria, and representatives of the genera sulfobacillus (iron- and sulfur-oxidizing firmicutes) and Oscillochloris (green sulfur bacteria) (Whitmarsh 1999). The Calvin cycle occurs in three stages as shown in Fig. 2.10: carboxylation of RuBP, reduction of 3-phosphoglycerate, and regeneration of RuBP.

2.4.1.1 Carboxylation of Ribulose Bisphosphate

The CO_2 molecules enter in the cycle by reacting with CO_2 acceptor RuBP to yield the first stable intermediate of the cycle, two molecules of 3-phosphoglycerate (3-PGA), reaction catalyzed by the enzyme RuBisCO (Kirk 1994). It is this 3-C molecule, the first stable product of the carboxylation reaction of RuBisCO that gives the C_3 cycle its name. The affinity of RuBisCO for CO_2 is sufficiently high to ensure rapid carboxylation at the low concentrations of CO_2 found in photosynthetic cells.

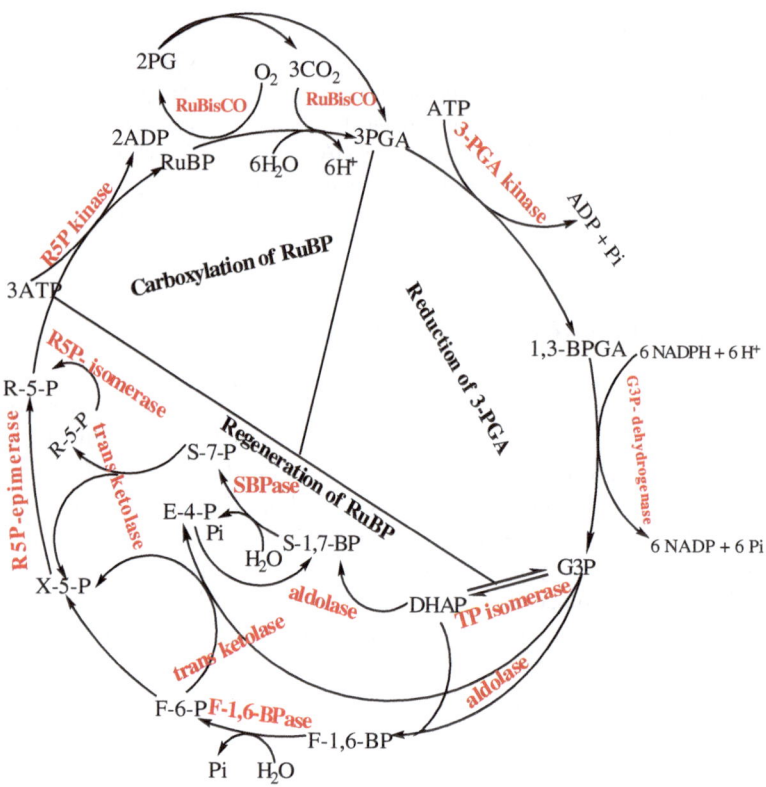

Fig. 2.10   Stages of Calvin cycle

2.4.1.2   Reduction of 3-Phosphoglycerate

The 3-phosphoglycerate formed in the carboxylation stage undergoes reductive phase of the cycle follows with two modifications:

i. The 3-PGA is first phosphorylated catalyzed by 3-PGA kinase to 1,3-bisphos-phoglycerate (1,3-BPGA) through use of assimilatory power, ATP, produced in the light reactions.
ii. Further, 1,3-BPGA is reduced to glyceraldehyde-3-phosphate (G-3-P), i.e., carboxyl group is transformed into an aldehyde group, through utilization of the NADPH produced by the light reactions, enzyme NADP: glyceraldehyde-3-phosphate dehydrogenase (GAPDH) catalyzes this step.

2.4.1.3   Regeneration of Ribulose-1,5-Bisphosphate

In order to keep the cycle functioning in the fixation of CO_2 in C_3 pathway, it is essential that necessary biochemical intermediates; CO_2 acceptor, RuBP, etc., be constantly regenerated. To prevent depletion of RuBP, G-3-P enters in the various sequential reactions of regenerative phase to ensure adequate supply of CO_2 acceptor. All steps of regeneration are summarized as follows:

1. One molecule of G-3-P is converted with the action of triose phosphate isomerase (TP isomerase) to dihydroxy-acetone-3-phosphate (DHAP) in an isomerization reaction.
2. DHAP then undergoes aldol condensation with a second molecule of G-3-P, a gluconeogenesis reaction catalyzed by aldolase to form fructose-1,6-bisphosphate (F-1,6-BP).
3. F-1,6-BP occupies a key position in the regeneration cycle and is hydrolyzed using enzyme fructose-1,6-bisphosphatase (F-1,6-BPase) to fructose-6-phosphate (F-6-P), which then reacts with the enzyme transketolase.
4. A 2-C unit (from C_1 and C_2 position) of donor F-6-P is transferred via transketolase to a third molecule of G-3-P acceptor to give erythrose-4-phosphate (E-4-P) and xylulose-5-phosphate (X-5-P).
5. Further, another aldol condensation via aldolase occurs between E-4-P and a fourth molecule of triose phosphate to yield the 7-C sugar sedoheptulose-1,7-bisphosphate (S-1,7-BP).
6. The S-1,7-BP is sequentially hydrolyzed by dephosphorylation at the C_1 position with the help of phosphatase (SBPase) to give sedoheptulose-7-phosphate (S-7-P).
7. The S-7-P donates a 2-C unit to the last 5th one molecule of G-3-P via transketolase enzyme and forms ribose-5-phosphate (R-5-P) and xylulose-5-phosphate (X-5-P).
8. The two molecules of X-5-P epimerized the C_3 position to form two molecules of R-5-P sugars by enzyme ribulose-5-phosphate epimerase (R5P-epimerase). The third molecule of R-5-P is isomerized to form R-5-P by ribose-5-phosphate isomerase (R5P-isomerase).
9. Finally, ribulose-5-phosphate kinase (R-5-P kinase) catalyzes the phosphorylation at the C_1 position of R-5-P with ATP thus regenerating the three essential molecules of the initial CO_2 acceptor, RuBP (total 15-C) by reactions that reshuffle the carbons from the five molecules of triose phosphate ($5 \times 3 = 15$ C).

Thus, the entire C_3 cycle consumes two molecules of NADPH and three molecules of ATP for every molecule of CO_2 fixed into hexose or carbohydrate (Falkowski and Raven 1997).

2.4.2 C_4 Cycle or Hatch–Slack Pathway

Apart from classical C_3 pathway, Hatch and Slack (1966) reported a new C_4 photosynthesis pathway of CO_2 fixation in sugar cane, which gives an initial four-carbon compound, oxalacetate, rather than 3-PGA. Although timing is uncertain, it is currently thought that C_4 pathway evolved gradually from C_3 ancestors (30 million years ago) through structural and biochemical modifications, in relation to environmental pressures (e.g., slumping ambient CO_2 level) (Xu et al. 2012). In order to evade the waste photorespiration pathway, C_4 plants possess an additional cytosolic carbon-fixing enzyme phosphoenolpyruvate carboxylase (PEP carboxylase) in addition to RuBisCO (Furbank and Taylor 1995; Chollet et al. 1996). The PEP carboxylase has a higher affinity for CO_2 (lower K_m) and lower affinity for O_2 (higher K_m) than RuBisCO (Sage 2002). Thus, PEP carboxylase is able to fix CO_2 at relatively low intracellular CO_2 concentration. C_4 plants are believed to have evolved gradually from C_3 plants through several intermediate stages of C_3–C_4 plants (Xu et al. 2012). It was also noted that C_4 plants usually show only small or even no apparent CO_2 loss in light. The explanation lies in their unique anatomy (called Kranz anatomy) (Kennedy 1976) and multiple carboxylation reactions. The C_4 terrestrial plants use a ring of specialized cells, bundle sheath cells, for more efficient C_3 carbon fixation, equipped with a CO_2-concentrating mechanism that supports carboxylation of ribulose-1,5-bisphosphate over oxygenation reactions.

As shown in Fig. 2.11, the primary step of C_4 cycle is the production of 3 carbon phosphoenolpyruvate (PEP) from pyruvate using enzyme pyruvate orthophosphate dikinase (PPDK, E.C. 2.7.9.1), inorganic phosphate, and ATP. The next step is the CO_2 fixation through irreversible β-carboxylation of phosphoenolpyruvate (PEP) to 4-carbon oxaloacetate (first stable product) in the presence of ubiquitous

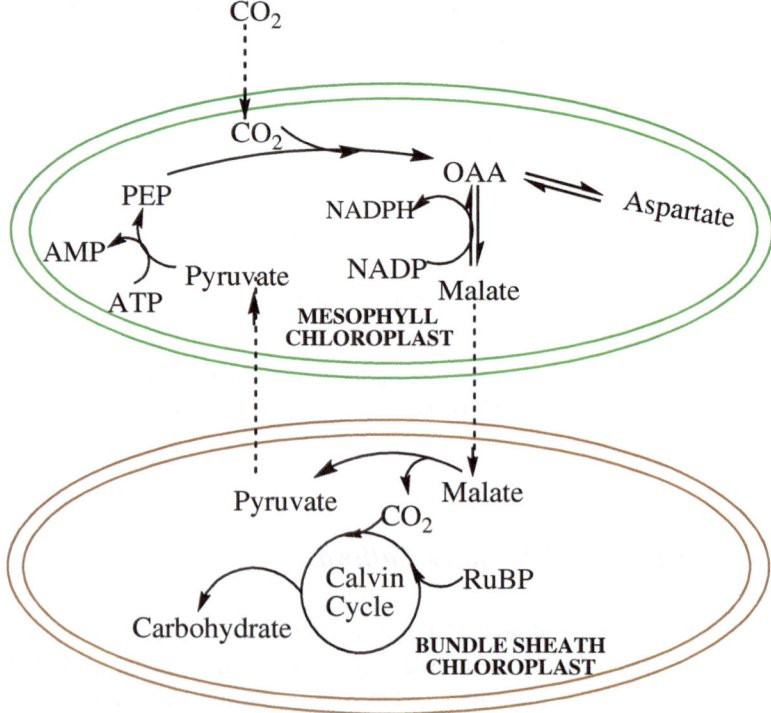

Fig. 2.11 Schematic representation of Hatch–slack pathway

enzyme PEP carboxylase, bicarbonate ions, and Me^{2+} (Furbank and Taylor 1995). Both steps occur in the mesophyll cells. Further, oxaloacetate can be reduced quickly to four-carbon malate using enzyme malate dehydrogenase in the leaf mesophyll cells. Malate is easily transferred to bundle sheath cells and converted to 3 carbon pyruvate, releasing CO_2 and reducing NADP to NADPH for C_3 cycle. This maintains CO_2 concentration high, so that RuBisCO is used almost entirely as a carboxylase, minimizing photorespiration. Thus, PEP carboxylase activity helps to refix any respired CO_2 formed from the oxygenase function of RuBisCO to prevent the photorespiratory release of CO_2.

Although, surprisingly lacks of Kranz dual-cell compartmentation in aquatic photosynthetic microorganisms, recent metabolic labeling and genome sequencing data suggest that the some species including green alga (*Chara contraria, Ostreococcus tauri,* and *Micromonas*) and diatoms (*Thalassiosira weissflogii, Phaeodactylum tricornutum,* and *T. pseudonana*) could also use the maneuver of both C_3 and C_4 fixation (Keeley et al. 1986; Keeley 1998; Derelle et al. 2006; Haimovichae Dayan et al. 2013). Presence of C_4 fixation has been further strengthened by the occurrence of relevant genes in their genomes. *Ostreococcus* has all the machinery; PEP carboxylase, $NADP^+$-dependent malic enzyme, and pyruvate orthophosphate dikinase with a predicted chloroplast-targeted transit peptides in a later two, necessary to perform C_4 fixation (van Ooijen et al. 2012). Interestingly, only one member of the marine Chlorophyta, macroscopic green macroalga *Udotea flabellum,* has been shown to perform C_4 photosynthesis (Reiskind and Bowes 1991). It shows that *U. flabellum* utilizes PEP carboxykinase (PEPCK) (as $NADP^+$-malic enzyme being absent), which activity in *Udotea* extracts is equivalent to RuBP carboxylase (Reiskind et al. 1988; Reiskind and Bowes 1991). Recently, the coexistence of genes necessary for both C_3 and C_4 pathway enzymes has also been reported in another green-tide-forming macroalga, *Ulva prolifera* (Xu et al. 2012). The expression levels of C_3 and C_4 photosynthesis genes, *rbc*L and PPDK, increased under stress conditions (such as high and low salinity, high and low temperature). However, contradictory experimental facts of Kremer and Kuppers (1977) shedded doubt on C_4 photosynthesis in algae. They investigated short-term (2–5 s) photosynthesis using $H^{14}CO_3^-$ in various species of different algal classes. *Ulva* produces malate and aspartate organic acids only as a minor component of short time ^{14}C-labeling fixation (less than 10 % of the total ^{14}C-labeling), while RuBP carboxylase (E.C. 4.1.1.39) was the main carbon-fixing enzyme. Usually, aquatic plants are subjected to much lower pCO_2 for photosynthesis. Thus, environmental stress may act as a major driving force to develop alterations of photosynthetic pathways toward C_4 metabolism for suppression of photorespiration. As another example, a submerged aquatic monocot plant *Hydrilla verticillata* operates a facultative, single-cell C_4 system (Rao et al. 2006), i.e., capable to change its photosynthetic pathway from C_3 to C_4 under conditions of CO_2 deficiency. However, studies of photosynthetic pathways of photosynthetic microorganisms are scanty, and there are extremely restricted knowledge of the pathways, which regulating the biology of the altered aquatic C_4 species. Thus, general occurrence of C_4-like mechanisms in aquatic photosynthetic microorganisms is therefore still in question. Despite its additional energetic cost, if photosynthetic microorganisms are capable of C_4 photosynthesis, it could comprise of a significant ecological benefit in CO_2-limiting conditions of phytoplankton blooms, particularly in conditions where competitors have inferior CCM efficiencies (or no CCM).

References

Alber BE, Ferry JG (1994) A carbonic anhydrase from the archaeon *Methanosarcina thermophila*. Proc Natl Acad Sci 91(15):6909–6913

Andersson I (2008) Catalysis and regulation in RuBisCO. J Exp Bot 59(7):1555–1568

Andersson I, Backlund A (2008) Structure and function of RuBisCO. Plant Physiol Biochem 46(3):275–291

Andersson I, Knight S, Schneider G, Lindqvist Y, Lundqvist T, Branden C-I, Lorimer GH (1989) Crystal structure of the active site of ribulose-bisphosphate carboxylase. Nature 337:229–234

Aspatwar A, Tolvanen MEE, Parkkila S (2010) Phylogeny and expression of carbonic anhydrase-related proteins. BMC Mol Biol 11(1):25

Badger MR, Bek EJ (2008) Multiple RuBisCO forms in proteobacteria: their functional significance in relation to CO_2 acquisition by the CBB cycle. J Exp Bot 59(7):1525–1541

Badger MR, Kaplan A, Berry JA (1980) Internal inorganic carbon pool of *Chlamydomonas reinhardtii*: Evidence for a carbon dioxide-concentrating mechanism. Plant Physiol 66(3):407–413

Badger MR, Price GD (1994) The role of carbonic anhydrase in photosynthesis. Annu Rev Plant Biol 45(1):369–392

Badger MR, Price GD (2003) CO_2 concentrating mechanisms in cyanobacteria: molecular components, their diversity and evolution. J Exp Bot 54(383):609–622

Barber J (2003) Photosystem II: the engine of life. Q Rev Biophys 36(01):71–89

Barsanti L, Gualtieri P (2014) Algae: anatomy, biochemistry, and biotechnology. CRC press, Boca Raton

Bartlett GJ, Porter CT, Borkakoti N, Thornton JM (2002) Analysis of catalytic residues in enzyme active sites. J Mol Biol 324(1):105–121

Berg JM, Tymoczko J, Stryer L (2002) Biochemistry making a fast reaction faster: carbonic anhydrases

Björn LO (2008) The evolution of photosynthesis and its environmental impact. In: Photobiology. Springer, pp 255–287

Blankenship RE (2014) Molecular mechanisms of photosynthesis. Wiley

Boone CD, Habibzadegan A, Gill S, McKenna R (2013) Carbonic anhydrases and their biotechnological applications. Biomolecules 3(3):553–562

Bracher A, Starling-Windhof A, Hartl FU, Hayer-Hartl M (2011) Crystal structure of a chaperone-bound assembly intermediate of form I RuBisCO. Nat Struct Mol Biol 18(8):875–880

Burnell JN, Gibbs MJ, Mason JG (1990) Spinach chloroplastic carbonic anhydrase nucleotide sequence analysis of cDNA. Plant Physiol 92(1):37–40

Calvin M, Bassham JA, Benson AA, Lynch V, Ouellet C, Schou L, Stepka Wt, Tolbert NE (1950) The path of carbon in photosynthesis. X. Carbon dioxide assimilation in plants. Lawrence Berkeley National Laboratory

Carrae-Mlouka A, Maejean A, Quillardet P, Ashida H, Saito Y, Yokota A, Callebaut I, Sekowska A, Dittmann E, Bouchier C (2006) A new RuBisCO-like protein coexists with a photosynthetic RuBisCO in the planktonic cyanobacteria *Microcystis*. J Biol Chem 281(34):24462–24471

Chollet R, Vidal J, O'Leary MH (1996) Phospho enol pyruvate carboxylase: a ubiquitous, highly regulated enzyme in plants. Annu Rev Plant Biol 47(1):273–298

Clegg MT, Cummings MP, Durbin ML (1997) The evolution of plant nuclear genes. Proc Natl Acad Sci 94(15):7791–7798

Cot SSW, So AKC, Espie GS (2008) A multiprotein bicarbonate dehydration complex essential to carboxysome function in cyanobacteria. J Bacteriol 190(3):936–945

Covarrubias AS, Bergfors T, Jones TA, Hogbom M (2006) Structural mechanics of the pH-dependent activity of beta-carbonic anhydrase from *Mycobacterium tuberculosis*. J Biol Chem 281(8):4993–4999

Daley SME, Kappell AD, Carrick MJ, Burnap RL (2012) Regulation of the cyanobacterial CO_2-concentrating mechanism involves internal sensing of $NADP^+$ and α-ketogutarate levels by transcription factor CcmR. PLoS ONE 7(7):e41286

Derelle E, Ferraz C, Rombauts S, Rouza P, Worden AZ, Robbens S, Partensky Fdr, Degroeve S, Echeynia S, Cooke R (2006) Genome analysis of the smallest free-living eukaryote Ostreococcus tauri unveils many unique features. Proc Nat Acad Sci 103(31):11647–11652

Donaldson TL, Quinn JA (1974) Kinetic constants determined from membrane transport measurements: carbonic anhydrase activity at high concentrations. Proc Natl Acad Sci 71(12):4995–4999

Ducat DC, Silver PA (2012) Improving carbon fixation pathways. Curr Opin Chem Biol 16(3):337–344

Dufossae L, Galaup P, Yaron A, Arad SM, Blanc P, Chidambara Murthy KN, Ravishankar GA (2005) Microorganisms and microalgae as sources of pigments for food use: a scientific oddity or an industrial reality? Trends Food Sci Technol 16(9):389–406

Ellis RJ (1979) The most abundant protein in the world. Trends Biochem Sci 4(11):241–244

Emlyn-Jones D, Woodger FJ, Price GD, Whitney SM (2006) RbcX can function as a RubBisCO chaperonin, but is non-essential in *Synechococcus* PCC7942. Plant Cell Physiol 47(12):1630–1640

Eriksson AE, Jones TA, Liljas A (1988) Refined structure of human carbonic anhydrase II at 2.0 Å resolution. Proteins Struct Funct Bioinf 4(4):274–282

Eriksson M, Karlsson J, Ramazanov Z, Gardestrom P, Samuelsson G (1996) Discovery of an algal mitochondrial carbonic anhydrase: molecular cloning and characterization of a low-CO_2-induced polypeptide in *Chlamydomonas reinhardtii*. Proc Natl Acad Sci 93(21):12031–12034

Falkowski PG, Raven JA (1997) Aquatic photosynthesis. Blackwell Science, New Jersey

Fawcett TW, Volokita M, Bartlett SG (1990) Spinach carbonic anhydrase primary structure deduced from the sequence of a cDNA clone. J Biol Chem 265(10):5414–5417

Ferguson JKW, Lewis L, Smith J (1937) The distribution of carbonic anhydrase in certain marine invertebrates. J Cell Comp Physiol 10(3):395–400

Flores-Paerez Ã, Jarvis P (2013) Molecular chaperone involvement in chloroplast protein import. Biochimica et Biophysica Acta (BBA)-Mol Cell Res 1833(2):332–340

Fukuzawa H, Fujiwara S, Yamamoto Y, Dionisio-Sese ML, Miyachi S (1990) cDNA cloning, sequence, and expression of carbonic anhydrase in *Chlamydomonas reinhardtii*: regulation by environmental CO_2 concentration. Proc Natl Acad Sci 87(11):4383–4387

Fukuzawa H, Suzuki E, Komukai Y, Miyachi S (1992) A gene homologous to chloroplast carbonic anhydrase (*icfA*) is essential to photosynthetic carbon dioxide fixation by *Synechococcus* PCC7942. Proc Natl Acad Sci 89(10):4437–4441

Furbank RT, Taylor WC (1995) Regulation of photosynthesis in C_3 and C_4 plants: a molecular approach. Plant Cell 7(7):797

Gilbert BC, Murphy DM, Chechik V (2012) Electron paramagnetic resonance. Royal Society of Chemistry, UK

Green B, Parson WW (2003) Light-harvesting antennas in photosynthesis. Springer, New York

Haimovichae Dayan M, Garfinkel N, Ewe D, Marcus Y, Gruber A, Wagner H, Kroth PG, Kaplan A (2013) The role of C_4 metabolism in the marine diatom *Phaeodactylum tricornutum*. New Phytol 197(1):177–185

Hanson TE, Tabita FR (2001) A ribulose-1, 5-bisphosphate carboxylase/oxygenase (RuBisCO)-like protein from *Chlorobium tepidum* that is involved with sulfur metabolism and the response to oxidative stress. Proc Natl Acad Sci 98(8):4397–4402

Hatch MD, Slack CR (1966) Photosynthesis by sugar-cane leaves. Biochem J 101:103–111

Heinhorst S, Williams EB, Cai F, Murin CD, Shively JM, Cannon GC (2006) Characterization of the carboxysomal carbonic anhydrase CsoSCA from *Halothiobacillus neapolitanus*. J Bacteriol 188(23):8087–8094

Hewett-Emmett D, Tashian RE (1996) Functional diversity, conservation, and convergence in the evolution of the α-, β-, and γ-carbonic anhydrase gene families. Mol Phylogenet Evol 5(1):50–77

Hiltonen T, Björkbacka H, Forsman C, Clarke AK, Samuelsson G (1998) Intracellular β-carbonic anhydrase of the unicellular green alga *Coccomyxa* cloning of the cDNA and characterization of the functional enzyme overexpressed in *Escherichia coli*. Plant Physiol 117(4):1341–1349

Huang S, Hainzl T, Grundstrom C, Forsman C, Samuelsson G, Sauer-Eriksson AE (2011) Structural studies of β-carbonic anhydrase from the green alga *Coccomyxa*: inhibitor complexes with anions and acetazolamide. PLoS ONE 6(12):e28458

Iverson TM, Alber BE, Kisker C, Ferry JG, Rees DC (2000) A closer look at the active site of γ-class carbonic anhydrases: high-resolution crystallographic studies of the carbonic anhydrase from *Methanosarcina thermophila*. Biochemistry 39(31):9222–9231

Kanth BK, Min K, Kumari S, Jeon H, Jin ES, Lee J, Pack SP (2012) Expression and characterization of codon-optimized carbonic anhydrase from *Dunaliella* species for CO_2 sequestration application. Appl Biochem Biotechnol 167(8):2341–2356

Karkehabadi S (2005) Structure-function studies of ribulose-1, 5-bisphosphate carboxylase/oxygenase: activation, thermostability, and CO_2/O_2 specificity, vol 2005

Keeley JE (1998) C_4 photosynthetic modifications in the evolutionary transition from land to water in aquatic grasses. Oecologia 116(1–2):85–97

Keeley JE, Sternberg LO, Deniro MJ (1986) The use of stable isotopes in the study of photosynthesis in freshwater plants. Aquat Bot 26:213–223

Kennedy RA (1976) Photorespiration in C_3 and C_4 plant tissue cultures significance of Kranz Anatomy to low photorespiration in C_4 Plants. Plant Physiol 58(4):573–575

Kimber MS, Pai EF (2000) The active site architecture of *Pisum sativum* β-carbonic anhydrase is a mirror image of that of α-carbonic anhydrases. EMBO J 19(7):1407–1418

Kirk JTO (1994) Light and photosynthesis in aquatic ecosystems. Cambridge University Press, Cambridge

Kisker C, Schindelin H, Alber BE, Ferry JG, Rees DC (1996) A left-hand beta-helix revealed by the crystal structure of a carbonic anhydrase from the archaeon *Methanosarcina thermophila*. The EMBO J 15(10):2323

Kremer BP, Kuppers U (1977) Carboxylating enzymes and pathway of photosynthetic carbon assimilation in different marine algae-Evidence for the C_4-pathway? Planta 133(2):191–196

Krishnamurthy VM, Kaufman GK, Urbach AR, Gitlin I, Gudiksen KL, Weibel DB, Whitesides GM (2008) Carbonic anhydrase as a model for biophysical and physical-organic studies of proteins and protein-ligand binding. Chem Rev 108(3):946–1051

Krissinel E, Henrick K (2007) Inference of macromolecular assemblies from crystalline state. J Mol Biol 372(3):774–797

Kröncke K-D, Klotz L-O (2009) Zinc fingers as biologic redox switches? Antioxid Redox Signal 11(5):1015–1027

Kumar S, Singh V, Tiwari M (2007) Quantitative structure activity relationship studies of sulfamide derivatives as carbonic anhydrase inhibitor: as antiglaucoma agents. Med Chem 3(4):379–386

Laing WA, Ogren WL, Hageman RH (1974) Regulation of soybean net photosynthetic CO_2 fixation by the interaction of CO_2, O_2, and ribulose 1, 5-diphosphate carboxylase. Plant Physiol 54(5):678–685

Lapointe M, MacKenzie TDB, Morse D (2008) An external delta-carbonic anhydrase in a free-living marine dinoflagellate may circumvent diffusion-limited carbon acquisition. Plant Physiol 147(3):1427–1436

Larimer FW, Soper TS (1993) Overproduction of *Anabaena* 7120 ribulose-bisphosphate carboxylase/oxygenase in *Escherichia coli*. Gene 126(1):85–92

Liljas A, Kannan KK, Bergsten PC, Waara I, Fridborg K, Strandberg B, Carlbom U, Järup L, Lagren S, Petef M (1972) Crystal structure of human carbonic anhydrase C. Nature 235(57):131–137

Liu C, Young AL, Starling-Windhof A, Bracher A, Saschenbrecker S, Rao BV, Rao KV, Berninghausen O, Mielke T, Hartl FU (2010) Coupled chaperone action in folding and assembly of hexadecameric RuBisco. Nature 463(7278):197–202

McGrath JM, Long SP (2014) Can the cyanobacterial carbon-concentrating mechanism increase photosynthesis in crop species? A theoretical analysis. Plant Physiol 164(4):2247–2261

Mitsuhashi S, Mizushima T, Yamashita E, Yamamoto M, Kumasaka T, Moriyama H, Ueki T, Miyachi S, Tsukihara T (2000) X-ray structure of beta-carbonic anhydrase from the red alga, *Porphyridium purpureum*, reveals a novel catalytic site for CO_2 hydration. J Biol Chem 275(8):5521–5526

Morita K, Hatanaka T, Misoo S, Fukayama H (2014) Unusual small subunit that is not expressed in photosynthetic cells alters the catalytic properties of RuBisco in rice. Plant Physiol 164(1):69–79

Moroney JV, Ma Y, Frey WD, Fusilier KA, Pham TT, Simms TA, DiMario RJ, Yang J, Mukherjee B (2011) The carbonic anhydrase isoforms of *Chlamydomonas reinhardtii*: intracellular location, expression, and physiological roles. Photosynth Res 109(1–3):133–149

Moroney JV, Ynalvez RA (2007) Proposed carbon dioxide concentrating mechanism in *Chlamydomonas reinhardtii*. Eukaryot Cell 6(8):1251–1259

Mukherjee B (2013) Investigation of the role of putative inorganic carbon transporters in the carbon dioxide concentrating mechanisms of *Chlamydomonas reinhardtii*. Calcutta University, Kolkata

Newman J, Gutteridge S (1993) The X-ray structure of *Synechococcus* ribulose-bisphosphate carboxylase/oxygenase-activated quaternary complex at 2.2-A resolution. J Biol Chem 268(34):25876–25886

Nishitani Y, Yoshida S, Fujihashi M, Kitagawa K, Doi T, Atomi H, Imanaka T, Miki K (2010) Structure-based catalytic optimization of a type III RuBisco from a hyperthermophile. J Biol Chem 285(50):39339–39347

Ogren WL (1984) Photorespiration: pathways, regulation, and modification. Annu Rev Plant Physiol 35(1):415–442

Onizuka T, Endo S, Akiyama H, Kanai S, Hirano M, Yokota A, Tanaka S, Miyasaka H (2004) The *rbcX* gene product promotes the production and assembly of ribulose-1, 5-bisphosphate carboxylase/oxygenase of *Synechococcus* sp. PCC7002 in *Escherichia coli*. Plant Cell Physiol 45(10):1390–1395

Parkin G (2004) Synthetic analogues relevant to the structure and function of zinc enzymes. Chem Rev 104(2):699–768

Pastorek J, Pastorekova S, Callebaut I, Mornon JP, Zelník V, Opavska R, Zat'ovicová M, Liao S, Portetelle D, Stanbridge EJ (1994) Cloning and characterization of MN, a human tumor-associated protein with a domain homologous to carbonic anhydrase and a putative helix-loop-helix DNA binding segment. Oncogene 9(10):2877–2888

Peña KL, Castel SE, de Araujo C, Espie GS, Kimber MS (2010) Structural basis of the oxidative activation of the carboxysomal gamma-carbonic anhydrase, CcmM. Proc Natl Acad Sci 107(6):2455–2460

Portis AR Jr (1992) Regulation of ribulose 1, 5-bisphosphate carboxylase/oxygenase activity. Annu Rev Plant Biol 43(1):415–437

Portis AR Jr (2003) RuBisco activase "RuBisco's catalytic chaperone". Photosynth Res 75(1):11–27

Portis AR Jr, Parry MAJ (2007) Discoveries in RuBisco (Ribulose 1, 5-bisphosphate carboxylase/oxygenase): a historical perspective. Photosynth Res 94(1):121–143

Price GD (2011) Inorganic carbon transporters of the cyanobacterial CO_2 concentrating mechanism. Photosynth Res 109(1–3):47–57

Price GD, Badger MR, Woodger FJ, Long BM (2008) Advances in understanding the cyano-bacterial CO_2-concentrating-mechanism (CCM): functional components, C_i transport-ers, diversity, genetic regulation and prospects for engineering into plants. J Exp Bot 59(7):1441–1461

Rae BD, Long BM, Badger MR, Price GD (2013) Functions, compositions, and evolution of the two types of carboxysomes: polyhedral microcompartments that facilitate CO_2 fixation in cyanobacteria and some proteobacteria. Microbiol Mol Biol Rev 77(3):357–379

Raines CA (2011) Increasing photosynthetic carbon assimilation in C_3 plants to improve crop yield: current and future strategies. Plant Physiol 155(1):36–42

Rao SK, Fukayama H, Reiskind JB, Miyao M, Bowes G (2006) Identification of C_4 responsive genes in the facultative C_4 plant *Hydrilla verticillata*. Photosynth Res 88(2):173–183

Raven JA (1997) CO_2-concentrating mechanisms: a direct role for thylakoid lumen acidification? Plant Cell Environ 20(2):147–154

Raven JA, Cockell CS, De La Rocha CL (2008) The evolution of inorganic carbon concentrating mechanisms in photosynthesis. Philos Trans R Soc B Biol Sci 363(1504):2641–2650

Reinfelder JR (2011) Carbon concentrating mechanisms in eukaryotic marine phytoplankton. Ann Rev Mar Sci 3:291–315

Reiskind JB, Bowes G (1991) The role of phosphoenolpyruvate carboxykinase in a marine mac-roalga with C_4-like photosynthetic characteristics. Proc Natl Acad Sci 88(7):2883–2887

Reiskind JB, Seamon PT, Bowes G (1988) Alternative methods of photosynthetic carbon assimi-lation in marine macroalgae. Plant Physiol 87(3):686–692

Roberts SB, Lane TW, Morel FoMM (1997) Carbonic anhydrase in the marine diatom *Thalassiosira weissflogii* (Bacillariophyceae) 1. J Phycol 33(5):845–850

Rudi K, Skulberg OM, Jakobsen KS (1998) Evolution of cyanobacteria by exchange of genetic material among phyletically related strains. J Bacteriol 180(13):3453–3461

Sage RF (2002) Variation in the k_{cat} of RuBisco in C_3 and C_4 plants and some implications for photosynthetic performance at high and low temperature. J Exp Bot 53(369):609–620

Saschenbrecker S (2007) Folding and assembly of RuBisCO. Ph.D. thesis, lmu

Saschenbrecker S, Bracher A, Rao KV, Rao BV, Hartl FU, Hayer-Hartl M (2007) Structure and function of RbcX, an assembly chaperone for hexadecameric RuBisco. Cell 129(6):1189–1200

Sawaya MR, Cannon GC, Heinhorst S, Tanaka S, Williams EB, Yeates TO, Kerfeld CA (2006) The structure of β-carbonic anhydrase from the carboxysomal shell reveals a distinct sub-class with one active site for the price of two. J Biol Chem 281(11):7546–7555

Schlicker C, Hall RA, Vullo D, Middelhaufe S, Gertz M, Supuran CT, Mahlschlegel FA, Steegborn C (2009) Structure and inhibition of the CO_2-sensing carbonic anhydrase Can2 from the pathogenic fungus *Cryptococcus neoformans*. J Mol Biol 385(4):1207–1220

Schneider G, Lindqvist Y, Lundqvist T (1990) Crystallographic refinement and structure of ribu-lose-1, 5-bisphosphate carboxylase from *Rhodospirillum rubrum* at 1.7 Å resolution. J Mol Biol 211(4):989–1008

Scholes GD, Fleming GR (2005) Energy transfer and photosynthetic light harvesting. Adv Chem Phys 132:57–130

Schuster G, Owens GC, Cohen Y, Ohad I (1984) Thylakoid polypeptide composition and light-independent phosphorylation of the chlorophyll *a*, *b*-protein in *Prochloron*, a prokary-ote exhibiting oxygenic photosynthesis. Biochimica et Biophysica Acta (BBA)-Bioenerg 767(3):596–605

Shiba T, Simidu U, Taga N (1979) Distribution of aerobic bacteria which contain bacteriochloro-phyll a. Appl Environ Microbiol 38(1):43–45

Smith KS, Ferry JG (1999) A plant-type (β-class) carbonic anhydrase in the thermophilic metha-noarchaeon *Methanobacterium thermoautotrophicum*. J Bacteriol 181(20):6247–6253

Smith MR, Mah RA (1978) Growth and methanogenesis by *Methanosarcina* strain 227 on ace-tate and methanol. Appl Environ Microbiol 36(6):870–879

So AKC, Espie GS (2005) Cyanobacterial carbonic anhydrases. Can J Bot 83(7):721–734

So AKC, Espie GS, Williams EB, Shively JM, Heinhorst S, Cannon GC (2004) A novel evolutionary lineage of carbonic anhydrase (ε-class) is a component of the carboxysome shell. J Bacteriol 186(3):623–630

Soltes-Rak E, Mulligan ME, Coleman JR (1997) Identification and characterization of a gene encoding a vertebrate-type carbonic anhydrase in cyanobacteria. J Bacteriol 179(3):769–774

Spreitzer RJ, Salvucci ME (2002) RuBisCO: structure, regulatory interactions, and possibilities for a better enzyme. Annu Rev Plant Biol 53(1):449–475

Sugawara H, Yamamoto H, Shibata N, Inoue T, Okada S, Miyake C, Yokota A, Kai Y (1999) Crystal structure of carboxylase reaction-oriented ribulose 1, 5-bisphosphate carboxylase/oxygenase from a thermophilic red alga, *Galdieria partita*. J Biol Chem 274(22):15655–15661

Sugiyama T, Mizuno M, Hayashi M (1984) Partitioning of nitrogen among ribulose-1, 5-bisphosphate carboxylase/oxygenase, phosphoenolpyruvate carboxylase, and pyruvate orthophosphate dikinase as related to biomass productivity in maize seedlings. Plant Physiol 75(3):665–669

Supuran CT (2011) Carbonic anhydrase inhibition with natural products: novel chemotypes and inhibition mechanisms. Mol Divers 15(2):305–16. doi:101007/s11030-010-9271-4 (Epub 2010 Aug 28)

Supuran CT, Scozzafava A (2007) Carbonic anhydrases as targets for medicinal chemistry. Bioorg Med Chem 15(13):4336–4350

Tabita FR (1999) Microbial ribulose 1, 5-bisphosphate carboxylase/oxygenase: a different perspective. Photosynth Res 60(1):1–28

Tabita FR, Hanson TE, Li H, Satagopan S, Singh J, Chan S (2007) Function, structure, and evolution of the RuBisCO-like proteins and their RuBisCO homologs. Microbiol Mol Biol Rev 71(4):576–599

Tabita FR, Satagopan S, Hanson TE, Kreel NE, Scott SS (2008) Distinct form I, II, III, and IV RuBisCO proteins from the three kingdoms of life provide clues about RuBisCO evolution and structure/function relationships. J Exp Bot 59(7):1515–1524

Taylor TC, Andersson I (1997) The structure of the complex between RuBisCO and its natural substrate ribulose 1, 5-bisphosphate. J Mol Biol 265(4):432–444

Tetu SG, Tanz SK, Vella N, Burnell JN, Ludwig M (2007) The Flaveria bidentis β-carbonic anhydrase gene family encodes cytosolic and chloroplastic isoforms demonstrating distinct organ-specific expression patterns. Plant Physiol 144(3):1316–1327

Tripp BC, Smith K, Ferry JG (2001) Carbonic anhydrase: new insights for an ancient enzyme. J Biol Chem 276(52):48615–48618

Tsuzuki M, Miyachi S (1989) The function of carbonic anhydrase in aquatic photosynthesis. Aquat Bot 34(1):85–104

Van Gorkom HJ (1985) Electron transfer in photosystem II. Photosynth Res 6(2):97–112

van Lun M, Hub JS, van der Spoel D, Andersson I (2014) CO_2 and O_2 distribution in RuBisco Suggests the small subunit functions as a CO_2 reservoir. J Am Chem Soc 136(8):3165–3171

van Ooijen G, Knox K, Kis K, Bouget F-Y, Millar AJ (2012) Genomic transformation of the picoeukaryote *Ostreococcus tauri*. J Vis Exp JoVE (65)

Vega MC, Lorentzen E, Linden A, Wilmanns M (2003) Evolutionary markers in the $(\beta/\alpha)8$-barrel fold. Curr Opin Chem Biol 7(6):694–701

Walker JCG (1985) Carbon dioxide on the early Earth. Orig Life Evol Biosph 16(2):117–127

Wang Q, Fristedt R, Yu X, Chen Z, Liu H, Lee Y, Guo H, Merchant SS, Lin C (2012) The γ-carbonic anhydrase subcomplex of mitochondrial complex I is essential for development and important for photomorphogenesis of *Arabidopsis*. Plant Physiol 160(3):1373–1383

Wang X, Wu S, Xu D, Xie D, Guo H (2011) Inhibitor and substrate binding by angiotensin-converting enzyme: quantum mechanical/molecular mechanical molecular dynamics studies. J Chem Inf Model 23; 51(5):1074–1082. doi:101021/ci200083f (Epub 2011 Apr 26)

Warlick B (2013) Functional discovery and promiscuity in the RuBisCO superfamily. University of Illinois at Urbana-Champaign, USA

Whitmarsh J (1999) The photosynthetic process. In: Concepts in photobiology. Springer, pp 11–51

Whitney SM, Andrews TJ (2001) The gene for the ribulose-1, 5-bisphosphate carboxylase/oxygenase (RuBisco) small subunit relocated to the plastid genome of tobacco directs the synthesis of small subunits that assemble into RuBisCO. Plant Cell Online 13(1):193–205

Whitney SM, Houtz RL, Alonso H (2011) Advancing our understanding and capacity to engineer nature's CO_2-sequestering enzyme, RuBisCO. Plant physiol 155(1):27–35

Windhof A (2011) RuBisco folding and oligomeric assembly: detailed analysis of an assembly intermediate. lmu

Xu J, Fan X, Zhang X, Xu D, Mou S, Cao S, Zheng Z, Miao J, Ye N (2012) Evidence of coexistence of C_3 and C_4 photosynthetic pathways in a green-tide-forming alga, *Ulva prolifera*. PloS one 7(5):e37438

Yurkov V, Beatty JT (1998) Isolation of aerobic anoxygenic photosynthetic bacteria from black smoker plume waters of the juan de fuca ridge in the pacific ocean. Appl Environ Microbiol 64(1):337–341

Zastrow ML, Pecoraro VL (2013) Influence of active site location on catalytic activity in de novo-designed zinc metalloenzymes. J Am Chem Soc 135(15):5895–5903

Chapter 3
Carbon-Concentrating Mechanism of Cyanobacteria

3.1 Introduction

Cyanobacteria, previously known as blue green algae, are a diverse group of Gram-negative oxygenic photosynthetic prokaryotes. On the basis of fossil record interpretations, phylogenetic, and biomarker investigations, they are thought to be appeared on earth approximately 2,600–3,500 million years ago (Rasmussen et al. 2008). The metabolic flexibility of cyanobacteria has enabled them to flourish under a wide range of niches such as, alkaline and acidic, hot and cold, marine, saline, terrestrial, freshwater, and symbiotic environmental conditions (Chorus and Bartram 1999; Whitehead 2013). Cyanobacteria are extremely productive, particularly open ocean species, which contribute about 30 % of worldwide net primary fixation (Rae et al. 2013a, b). However, this productivity would not be possible without carbon-concentrating mechanism (CCM) system which helps them to perform oxygenic photosynthesis and respiration within the same cell and allowed to generate energy molecules from diverse sources. Over the past decade, immense strides have been made in understanding the molecular details of the CCM in cyanobacteria (Badger and Price 2003; Pena et al. 2010; Rae et al. 2013a, b; Sandrini et al. 2014). Cyanobacterial CCMs feature multiple inorganic carbon (principally HCO_3^- and CO_2 species in aquatic environment) transport systems for the active uptake of inorganic carbon (C_i). The transported C_i further collected into the cytoplasm of cell as a pool of the relatively membrane-impermeable species, HCO_3^-. Consequently, this pool is converted to CO_2 via carboxysomal carbonic anhydrase (CA) inside a specialized microcompartment, carboxysome, in which it is fixed into 3-PGA by CO_2-fixing enzyme RuBisCO. This chapter provides the basic information related to the operation of cyanobacterial CCM in the view of latest findings.

© The Author(s) 2014
S.K. Singh et al., *Photosynthetic Microorganisms*, SpringerBriefs in Materials,
DOI 10.1007/978-3-319-09123-5_3

3.2 Structure and Types of Cyanobacteria

3.2.1 Morphological Features

The word "cyanobacteria" means blue color of the bacteria (Greek: kyanós = blue). These prokaryotic algal forms are more strongly associated with prokaryotic bacteria than eukaryotic algae. As like prokaryotic life form, they do not have any interior membrane organization or membrane-bound organelles (see Fig. 3.1). Similar to Gram-negative bacteria, cyanobacterial cell membrane has an outer lipid-rich peptidoglycan layer, composed of two sugar derivatives—N-acetylglucosamine (NAG) and N-acetylmuramic acid (NAM)—and several different amino acids (Hoiczyk and Hansel 2000; Wang and Chen 2009). However, cyanobacterial peptidoglycan layer is significantly broader than that of most Gram-negative bacteria; for example, peptidoglycan layer thickness of unicellular *Synechococcus lividus* (Golecki 1979) is about 10 nm, reached 15–35 nm in filamentous *Phormidium uncinatum* species, and increases more than 700 nm in large cyanobacteria like *Oscillatoria princeps* (Hoiczyk and Baumeister 1995; Hoiczyk and Hansel 2000). The peripheral protoplasm of cyanobacteria is primarily made of thylakoids with water-soluble chromo proteins phycobilisomes (phycobiliproteins) and glycogen granules (Colyer et al. 2005). The phycobilisome components (phycobiliproteins) are responsible for the blue green pigmentation of most cyanobacteria (Nordlund 2011). However, light color and quality drastically changes the phycobilisomes composition and in order to maximize the use of available light for photosynthesis, cyanobacteria adapted through complementary chromatic adaptation (CCA) process (Gutu et al. 2013). Thus, they appear green in red light (promotes phycocyanin-ll) and red in green light (accumulate more phycoerythrin). The 70S ribosomes are also dispersed throughout

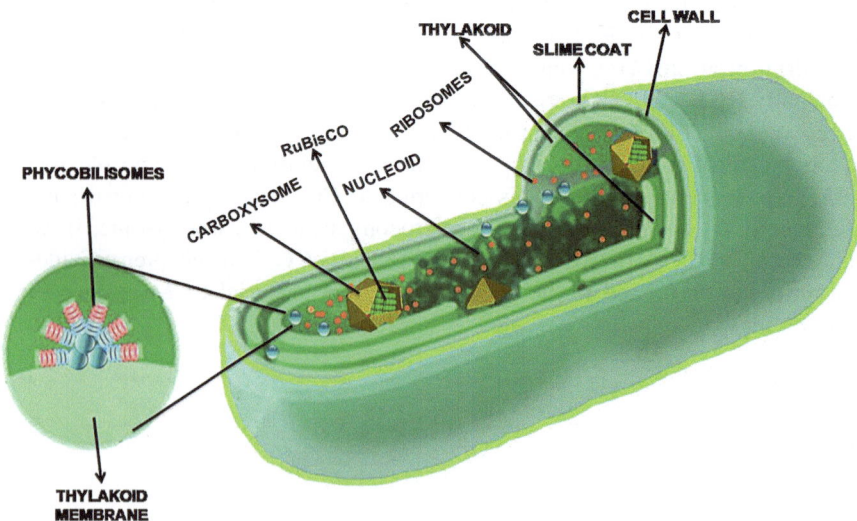

Fig. 3.1 Diagrammatic representation of cyanobacterial structure

the cell, but highest density occurred in the innermost area around the nucleoplasm (Lee 2008). Thylakoid membrane attached phycobilisomes and chlorophyll a; both act as major components of the photosynthetic light-harvesting antennae for the photosystems. However, few cyanobacteria lack phycobilisomes and have chlorophyll b (*Prochlorococcus*, *Prochlorothrix*, *Prochloron*) and chlorophyll d (*Acaryochloris marina*) (Chen et al. 2005). Several cyanobacterial species also having gas vacuoles help to adjust their buoyancy and stay close to the surface of water. Some cyanobacteria are also capable of gliding with two additional layers outside of the cell wall: serrated external layer (S-layer) and oscillin outside the outer membrane (Hoiczyk and Hansel 2000). In the cytoplasm, a temporary membraneless nitrogen reserve granules, non-ribosomally synthesized protein-like polymer "cyanophycin" or multi-l-arginyl-poly(l-aspartic acid) also occurs in nitrogen-fixing cyanobacteria (Steinle et al. 2008). When nitrogen-fixing cyanobacteria are starved for nitrogen, cyanophycin is broken down, while nitrogen is stored in phycobilisomes in cyanobacteria that do not fix nitrogen. In addition, an extraordinarily efficient proteinaceous organelle "carboxysomes" also present in all cyanobacterial strains and contain CO_2-fixing enzyme ribulose-1,5-bisphosphate carboxylase/oxygenase (RuBisCO).

3.2.2 Types

From the viewpoint of C_i acquisition, RuBisCO, and carboxysome lineages, cyanobacteria can be divided into two major groups: α-cyanobacteria and β-cyanobacteria. The α-cyanobacteria possess Form 1A RuBisCO within α-carboxysomes, while Form 1B RuBisCO associated with β-carboxysomes in β-cyanobacteria (Zarzycki et al. 2013). Both species are proteomically different in the carboxysomes and occupy a quite different range of ecological habitats (Price et al. 2008). But, each type is ultrastructurally similar. The α-cyanobacteria are tiny cells (1.6–2.8 Mb genome size), living in very stable open ocean environments with high pH (8.2) (Badger et al. 2006) and low-light–low-nutrient conditions. Although they have been illustrated as short flux, slow growers, low-energy cyanobacteria with a "minimal" CCM (Whitehead et al. 2014), they are highly abundant and productive organisms in open ocean environments. Thus, α-cyanobacteria could contribute as much as half of oceanic primary productivity (Whitehead et al. 2014); that is, up to 25 % net global primary productivity occurs in the oceans every year (Price et al. 2013). In contrast, β-cyanobacteria are characterized by larger cells (2.2–3.6 Mb genome size), relatively much more diverse range of habitats including freshwater, estuarine, hot springs, etc. (Whitehead et al. 2014). However, they nevertheless reach the α-cyanobacteria kind of global abundance (Badger et al. 2006). In addition to these groups, few of the α-cyanobacteria have been known as "transitional" strains (Rae et al. 2012), e.g., *Cyanobium* PCC7001, *Synechococcus* WH5701, and *Cyanobium* PCC6307. Similar to α-cyanobacteria, their C_i acqution systems are more diverse but also share a number of characteristics of β-cyanobacteria. They live in marginal marine and freshwater environments (Whitehead et al. 2014). It is believed that during evolutionary branching, they acquired the additional genes for encoding the C_i transporters through swapping the genetic material (lateral gene transfer) from contemporary β-cyanobacterial strains (Rae et al. 2012).

3.3 Inorganic Carbon (C_i) Acquisition Systems

Molecular and biochemical evidence indicates that there are five different inorganic carbon (C_i) acquisition systems in cyanobacteria, including three transporters for HCO_3^- (at plasma membrane) and two uptake systems for CO_2 (at thylakoid membrane), each with unique affinity and uptake flux capacity (Daley et al. 2012) (see Fig. 3.2).

It is believed that all of the active C_i uptake systems in cyanobacteria became inactive in dark, perhaps to avoid futile cycling (pump and leak), a process that potentially involves protein phosphorylation events, but upon lighting, acquisition systems are rapidly activated (Price 2011). Although these C_i acquisition systems differ in working mechanism, they are categorized into two broad groups: higher-affinity/low-flux (activate in low C_i) and lower-affinity/high-flux (activate in high C_i) systems (Daley et al. 2012). Table 3.1 shows the various inorganic carbon (C_i)

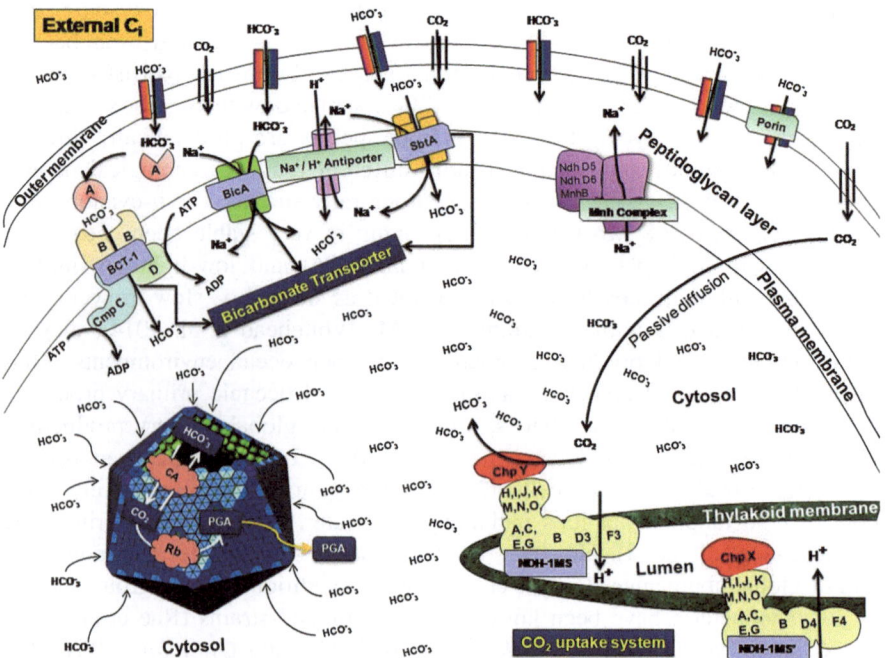

Fig. 3.2 Inorganic carbon (C_i) acquisition systems of cyanobacteria

Table 3.1 Presence and absence of inorganic carbon (C$_i$) uptake systems in α and β cyanobacterial strains

	Strains	Genes →	Bicarbonate transporter			uptake system		Carboxysomal proteins			
			cmpABCD	sbtA	bicA	ndhD$_3$F$_3$ CupAS	ndhD4F4 CupB	CcmK2K1 LMNO	Ccm K$_3$K$_4$	CsoS2S3 ABCsoS1	IcfA (Cca)
α	PM- NATL2A										
	PM-AS9601										
	PM- SS120										
	PM- MED4										
	PM- MIT9312										
	PM-MIT-9215/ 9211/ 9301/9303/9313/9515										
	SCH-WH8102										
	SCH-WH7803										
	SCH-RCC307										
	SCH-CC-9605/9902										
β	SCH-PCC6301										
	SCH-PCC7002										
	SCH-PCC7942										
	CY- ATCC 51142										
	AP-NIES-39										
	GV -PCC 7421										
	TSE-BP-1										
	MA- NIES-843										
	SCY- PCC 6803										
	CW- WH 8501										
	TDE-IMS101										
	NP- ATCC 29423										
	A -PCC 7120										
	AV-ATCC 29413										

PM: Prochlorococcus marinus, SCH: Synechococcus, GV: Gloeobacter violaceus, TSE: Thermosynechococcus elongates, SCY: Synechocystis, CW: Crocosphaera wastonii, TDE: Trichodesmium erythraeum, NP: Nostoc punctiforme, A: Anabaena, AV: Anabaena variabilis, CY: Cyanothece sp., AP: Arthrospira platensis, MA: Microcystis aeruginosa

acquisition systems of different cyanobacterial strain, on the basis of data available on http://genome.microbedb.jp/CyanoBase.

3.3.1 Bicarbonate Uptake System

3.3.1.1 Bicarbonate Uniporter Membrane Protein

BCT1

The bicarbonate transporter (BCT1) is a high-affinity bicarbonate multimeric (four protein subunits) complex, belonging to the diverse subfamily of bacterial ATP-binding cassette (ABC) transporter (Higgins 2001). This ATP-dependent uniporter was the first cyanobacterial C$_i$ transporter convincingly identified and characterized in freshwater β-cyanobacterium *Synechococcus* sp. PCC7942 (Omata et al. 1999; Omata et al. 2002). It is encoded by the *cmpAB(porB)CD* (*slr0040-44*) operon and expressed under severe C$_i$ limitation (Price et al. 2008) (see Fig. 3.3).

As shown in Fig. 3.4, *cmpA* encodes a 42 kD, a high-affinity cytoplasmic solute-binding extrinsic lipoprotein, localized as an array of binding proteins at the periplasmic face of plasma membrane. The stoichiometry of *cmpA* is apparently high relative to the other intrinsic subunits of BCT 1 transporter (Omata and Ogawa 1986). It acts as a substrate-binding subunit of the complex, specifically mobile scavenger for bicarbonate ($K_d = 5$ μM) (Maeda et al. 2000). To facilitate bicarbonate scavenging, *cmpA* can diffuse laterally through the N-terminal extension of mature polypeptide attached to outer surface of plasma membrane. The

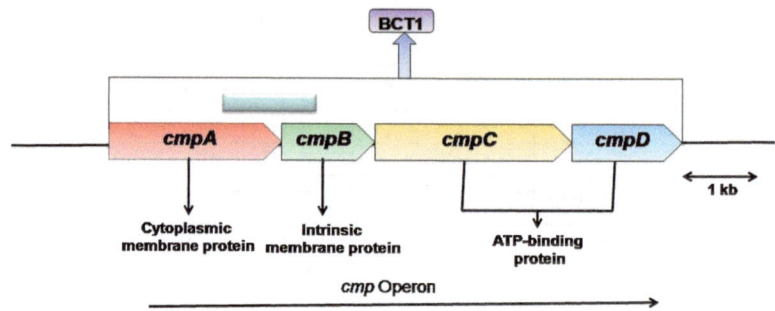

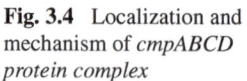

Fig. 3.3 Organization of *cmpABCD* gene cluster of *Synechococcus* sp. PCC 7942 strain

Fig. 3.4 Localization and mechanism of *cmpABCD* protein complex

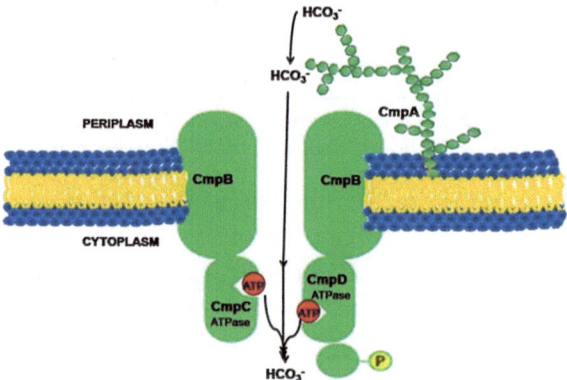

captured bicarbonate ions further transported to *cmpB*, an integral membrane permease. The *cmpB* subunits are dimerized hydrophobic intrinsic membrane polypeptide, which most probably forms a transport path within the membrane by analogy to other ABC transporters (Price 2011). The other both extrinsic membrane polypeptide subunits *cmpC* (an ATPase/solute-binding protein, large protein) and *cmpD* (a cytoplasmic ATPase, smaller protein) act as conserved ATP-binding sites and located on the cytoplasmic face of *cmpB* (Price et al. 2008).

In contrast to freshwater β-cyanobacteria, BCT1 are non-functional or not present in marine cyanobacterial strains (marine *Synechococcus* and *Prochlorococcus*) except *Synechococcus* sp. WH5701 (Rocap et al. 2002), as seawater is relatively rich in bicarbonate than freshwater (Badger et al. 2002) (see Table 3.1). Thus, it is possible that cyanobacteria acquired this transporter in the early stage of evolution and lost in marine strains during their evolutionary branching.

3.3.1.2 Sodium-Dependent Bicarbonate Transporter

There are two sodium-dependent bicarbonate transporters that have been identified in cyanobacterial strains and require Na^+ for HCO_3^- uptake (Herrero and Flores 2008). They are encoded by the genes, *bicA,* and *sbtA*, located on the same operon.

sbtA

The sodium-dependent bicarbonate (Na^+/HCO_3^-) transporter (*sbtA*) is a 160-kD multipass tetrameric membrane protein and a high-affinity ($K_{0.5} < 5$ μM HCO_3^-) symporter (*sbtA or slr1512 gene* encoded) (Seckbach et al. 2013), driven by an electrochemical Na^+ gradient (Price 2011). They were first identified in *Synechocystis* sp. strain PCC 6803 (Shibata et al. 2002). Interestingly, it works at low HCO_3^- and high Na^+ concentration. Dependency on Na^+ ions for bicarbonate transport helps to maintain the internal pH, by working mutually with Na^+/H^+ antiporter, during CO_2 fixation (Ogawa and Kaplan 2003). They have ten transmembrane spanning domains (374 amino acids, ≃39.7 kD) and two centrally located extrinsic hydrophilic regions (60 amino acids) that are possibly facing the cytoplasm (Badger and Price 2003). The *sbtA* homologs have been reported in many freshwater β-cyanobacterial genome (*Thioalkalivibrio* K90 mix, *T. thiocyanoxidans*), haloalkalitolerant cyanobacteria (*Synechococcus* sp., *Anabaena* sp., *Cyanothece* sp.), and alkaliphilc *Bacillus* (*Bacillus psudofermus, B. Halodurans*) (Seckbach et al. 2013, Badger and Price 2003). Nevertheless, weak homologs are reported in marine α-cyanobacterial species (Badger et al. 2002) (see Table 3.1). In addition to *sbtA*, Wang et al. (2004) reported another homologous small gene, *sbtB*, also at downstream of *sbtA* in *Synechocystis*, coexpressed with *sbtA* under C_i limitation.

bicA

The third bicarbonate transporter, *bicA*, is another Na^+-dependent HCO_3^- low-affinity transporter, first identified in the marine cyanobacterium *Synechococcus* PCC 7002 (Price et al. 2004). Homologs of *bicA* are identified widespread in both α- and β-cyanobacteria. It is potentially an important component of C_i uptake in many oceanic species of cyanobacteria. These transporters were also identified and characterized from *Synechococcus* WH 8102 and *Synechocystis* PCC6803 (Price et al. 2004, Xu et al. 2008) but not found in *Synechococcus* PCC7942 (Price et al. 2004). The expression studies of *bicA* showed that under C_i starvation, it is highly inducible in *Synechococcus* PCC 7002, while constitutively expressed in *Synechocystis* (Price et al. 2004, Wang et al. 2004). In contrast, recent study showed that freshwater cyanobacteria *Microcystis* strains with *bicA* performed better at high-C_i conditions (Sandrini et al. 2014). On the basis of phoA-lacZ topology fusion mapping, Shelden et al. (2010) predicted in *Synechococcus* PCC7002 that *bicA* could have up to 10–12 transmembrane domains in length of 566 amino acids (59.6 kD). It belongs to a large sulfate transporters or permeases family of eukaryotic and prokaryotic transporters often denoted as "SulP family" (Badger et al. 2006). The protein features a cytoplasmically located domain-designated STAS (sulfate transporter antisigma factor-like domain) domain, characteristic of the SulP family. The STAS domain of SulP transporters is also known to be a regulatory domain in some mammalian homologs (Ko et al. 2004).

3.3.2 Carbon Dioxide Uptake System

Although, it is conventionally accepted that CO_2 diffuse passively through water channels, such as aquaporin, to inside the cyanobacterial cell. But recent studies on type I NAD (P)H dehydrogenase (NDH-I) complex of cyanobacterial species provide new insights into functional mechanisms of CO_2 uptake system. Latest molecular, biochemical, and physiological studies have significantly confirmed their role in conversion of CO_2 to HCO_3^- and transport across thylakoid membrane, but the exact working mechanism still unclear.

3.3.2.1 NDH-I Complexes

The type I NAD(P)H dehydrogenase (NDH-I) complexes were first discovered in thylakoid and cytoplasmic membranes of cyanobacterium *Synechocystis* sp. PCC 6803 about 20 years ago by Berger and coresearchers (Berger et al. 1991). These proton-pumping complexes attracted much attention because of their vital dual physiological roles in chlororespiration as well as in NDH-I-dependent cyclic photophosphorylation pathways of photosynthesis in the light (Battchikova et al. 2011). Many studies revealed that variants of NDH-I complexes also participate in the uptake of CO_2 into the cell (Ohkawa et al. 2000, Shibata et al. 2001, Zhang et al. 2005, Bernat et al. 2011). However, the detailed arrangement and work mechanism of these complexes is still in its infancy. The depiction that is predicted from combination of structural biology, molecular biology, reverse genetics, and functional proteomics is that the cyanobacteria evolved different variants of NDH-I complex depending on the physiological needs of the cell. In cyanobacteria of *Synechocystis* PCC 6803, the CO_2 uptake is dependent on two variants of NDH-I complexes: NDH-IMS (Medium, ~350 kD) and NDH-IMS' (Small, ~200 kD). Other two variants, NDH-IL (Large, ~450 kD) and NDH-IL', are participated in respiration and cyclic electron flow during photosynthesis (see Fig. 3.5). However, basic core module of variants is composed of subcomplex NDH-IM which contains 14 subunits (*ndh*-ABCEGHIJKLMNOS). Besides the core module NDH-IM, NDH-IMS is functionally associated with other four subunits *ndh*D3 (54 kD), *ndh*F3 (66 kD), cupA (51 kD), and cupS which are essentially required in catalyzing active CO_2 uptake (Badger and Price 2003). *ndh*F and *ndh*D gene families (Battchikova et al. 2011; Battchikova and Aro 2007) may be related to an earliest gene duplication incident during evolutionary branching. The number of both gene families varies with cyanobacterial strain (Battchikova et al. 2005), e.g., *Synechocystis* PCC 6803 (three *ndh*F(F_1,F_3 and F_4) and six *ndh*D(D_1–D_6)), *Anabaena* sp. PCC 7120 (one *ndh*F(F_3) and one ndhD(D_3)), *Synechococcus* sp. WH8102 (two *ndh*F (F_1 and F_2), and one *ndh*D(D_1–D_3)) (Battchikova et al. 2010). NDH-IMS' is likely to include NDH-IM along with subunits *ndh*D4, *ndh*F4, and *cupB* (homolog of *cupA*, and *cupS*). It appears that this complex is constitutively expressed and shows low affinity to CO_2 (Zhang et al. 2004) in contrast to high-CO_2 (H-CO_2)-affinity NDH-IMS complex, which is induced under low

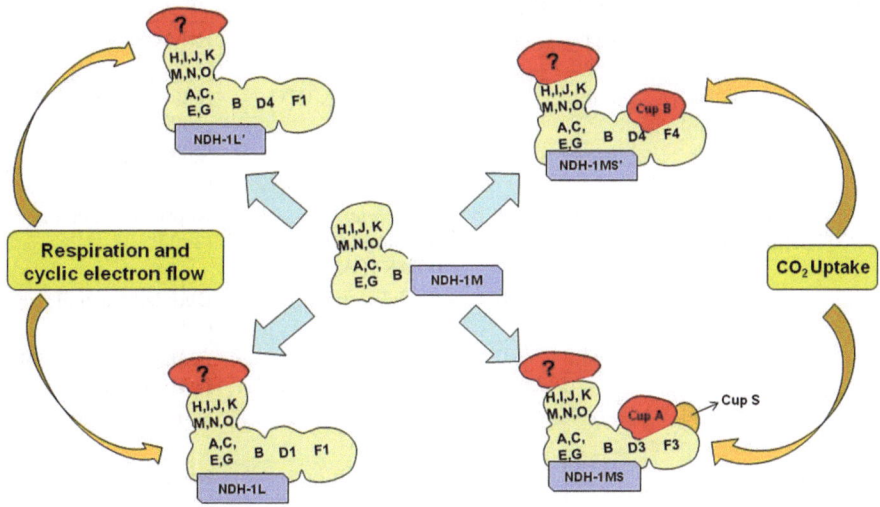

Fig. 3.5 Dual physiological role of NDH-1 protein complexes

CO$_2$ (L-CO$_2$) (Shibata et al. 2001). In addition to NDH-IM subunits, the NDH-IL complex contains *ndh*D1 and *ndh*F1 as well as two recently found subunits *ndh*P and *ndh*Q. The NDH-IL' complex is similar to NDH-IL except the fact that it has *ndh*D$_2$ instead of *ndh*D$_1$. Price et al. (2002) speculate that the water-soluble protein subunits cupA, cupS, and cupB are presumed to have a role in the conversion of CO$_2$ to HCO$_3^-$ in the cell. In *Synechocystis* 6803, H-CO$_2$ affinity NDH-IMS complex is strongly inducible in cyanobacterial cells grown photoautotrophically under L-CO$_2$ condition, but the expression is drastically reduced under rise in CO$_2$ levels (Herranen et al. 2004).

3.4 Cyanobacterial Carbonic Anhydrases

The first cyanobacterial CA was identified in cyanobacterium *Synechococcus* PCC7942 by two groups (Fukuzawa et al. 1992; Yu et al. 1992). Studies of the CCM in the cyanobacterium led to the simultaneous identification of wide diversity of CAs, α-CA being periplasmic, β-CA and γ-CA associated with the carboxysome (Kimber 2014) (see Table 3.2).

3.4.1 α CAs

Although the specific activity of α CA$_S$ has not yet been confirmed, it presumed to be present at the periplasmic space of some cyanobacterial strains. Soltes-Rak et al. (1993) reported the presence of external *α*-CA (*ecaA*) in two species

Table 3.2 Cyanobacterial carbonic anhydrases

CA class	Location	Gene	MW (kD)	Organism	References
α	Exreacellular periplasmic space	ecaA	29	Anabaena sp. PCC 7120	Soltes-Rak et al. (1993)
	Exreacellular periplasmic space	ecaA	26	Synechococcus elongates PCC 7942	Soltes-Rak et al. (1993)
	Exreacellular periplasmic space	ecaA	34	Microcoleus chthonoplastes— IPPAS B-353	Kupriyanova et al. (2007)
	Exreacellular periplasmic space	ecaA	60	Rhabdoderma lineare IPPAS B-354	Dudoladova et al. (2007)
β	Carboxysome	icfA	30	Synechococcus elongates PCC 7942	Fukuzawa et al. (1990), Price et al. (1992)
	Extracellular periplasmic space	ecaB	27.9	Synechocystis sp. PCC 6803	So et al. (1998)
	β-cyanobacterial carboxysomes	ccaA	30.7	Synechocystis sp. PCC 6803	So and Espie (1998)
	α-cyanobacterial carboxysomes	csoSCA	63.9	Synechocystis sp. WH 8102	Palenik et al. (2003)
	Exreacellular periplasmic space	cahB1	30.4	Microcoleus chthonoplastes - IPPAS B-353	Kupriyanova et al. (2011)
	Carboxysome		25	Rhabdoderma lineare IPPAS B-354	Dudoladova et al. (2007)
	Thylakoid membranes (PSII)		60	Rhabdoderma lineare IPPAS B-354	Dudoladova et al. (2007)
γ	β-cyanobacterial carboxysomes	ccmM	57.8	Synechococcus elongates PCC 7942	Price et al. (2002)

of cyanobacteria, *Anabaena* sp. PCC 7120 and *Synechococcus* sp. PCC 7942. They found that expression of *ecaA* in grown cells was regulated by CO_2 concentration and highest at elevated levels of CO_2. Further immunogold localization studies revealed the extracellular location of *ecaA* in the periplasmic space. Kupriyanova et al. (2007) also reported putative active extracellular α-CA (34 kD) in the alkaliphilic stromatolite-forming cyanobacterium *Microcoleus chthonoplastes*, localized in the glycocalyx of periplasm. It is believed that *ecaA* helps in the formation and maintenance of the equilibrium between C_i speciation (CO_2 and HCO_3^-) at the periplasmic site of transport to cytoplasmic transporters in order to maximize substrate availability. Alternatively, *ecaA* may also serve as a sensor (either detecting or signaling) of CO_2 level changes in the external medium. However, researchers found no detectable difference in CA activity of the mutant and the wild-type cells (Soltes-Rak et al. 1993; So et al. 1998). Thus, these extracellular CAs had very little effect on C_i uptake, and their contribution to the CCM was rather limited.

3.4.2 β-CAs

In cyanobacteria, on the basis of amino acid sequence similarities, several β CA proteins have been identified on extracellular (*ecaB*/*cahB1*) and carboxysomal (*icfA*/*ccaA, csoSCA*). The extracellular CA of the β class, *ecaB* (external CA beta class) gene, had been found in *Synechocystis* sp. PCC 6803 (So et al. 1998) which encodes a 263 amino acid polypeptide. The location of putative prokaryotic protein *ecaB* in *Synechocystis* PCC 6803 is still unclear. However, this protein had a binding site for a membrane lipoprotein, which indicates its possible location associated with the membrane or in the periplasmic space. Similar to *ecaA*, the enzyme activity of *ecaB* has also not been confirmed experimentally. It is possible that they are functional analogs, whose precise role in the CCM is still to be determined. One more carboxysomal representative of β CAs, *icfA*, had been isolated in *S. elongatus* PCC 7942 through complementation of a high-CO_2-requiring (5 %) mutant, in which carboxysomal CA activity was 10 times lower than the wild-type cells (Yu et al. 1992; Fukuzawa et al. 1992). Their results showed the putative CA encoded by *icfA* is essential to C_i fixation in cyanobacteria and proper functionality of carboxysome. The significant sequence similarities of *icfA* with plant chloroplast CAs have attributed to the fact that plant chloroplast CAs may have evolved from a common ancestor of the prokaryotic CAs. *icfA* homolog, *ccaA*, was found in *Synechocystis* sp. PCC 6803 (So and Espie 1998). Nowadays, *icfA* and *ccaA* enzymes have been shared a common class "carboxysomal carbonic anhydrase" (*ccaA*) (Price et al. 1998). This class of CA associated with the inner surface of the carboxysomal shells as a part of the multiprotein complex and converts HCO_3^- into CO_2 for fixation by RuBisCO. The *ccaA* gene has been found only in representatives of β-cyanobacteria; however, some of them do not possess such gene. The carboxysome shells of α-cyanobacteria carry another β CA, *csoSCA*, which performs the same function as ccaA of β cyanobacteria (Sawaya et al. 2006). Interestingly, Kupriyanova et al. (2011) characterized an another extracellular β class, *cahB1*, in alkaliphilic cyanobacterium *Microcoleus chthonoplastes* which is similar in amino acid and nucleotide sequences to β-CAs of *Synechococcus* sp. PCC 7942 (*icfA*) and *Synechocystis* sp. PCC 6803 (*CcaA*). *cahB1* possibly involves in the maintenance of a balance between external sources of C_i and *Microcoleus* photosynthesis in extreme environments of soda lakes.

3.4.3 γ-CAs

There has been only one γ CA, *ccmM* gene, characterized in all members of β cyanobacterial species. It encodes a bifunctional protein (70 kD) in *Synechococcus* PCC7942 (Cannon et al. 2010). The N-terminal domain is homologous to a γ-carbonic anhydrase-like domain from archaeon *Methanosarcina thermophila* (Cam) (Alber and Ferry 1994), binds with carboxysomal carbonic anhydrase

(CcaA), and converts bicarbonate into CO_2 to RuBisCO as a substrate for fixation. However, C-terminal domain of CcmM consists of 3–5 repeats of the small subunit of RuBisCO (*rbcL*)-like repeats (Cot et al. 2008, Long et al. 2010) and anchors the ccmM near the catalytic centers of RuBisCO bond with RuBisCO. It is believed that *CcmM* protein is a part of the *ccmKLMN* operon and essentially required for β carboxysomes assembly (Ogawa et al. 1994, Ludwig et al. 2000). Paradoxically, many β-cyanobacteria, including *Arthrospira platensis*-NIES-39, *Anabaena variabili* ATCC 29413, *Nostoc punctiforme* ATCC 29423, and *Thermosynechococcus elongatus* BP-1, lack the conventional carboxysomal β-CA, *ccaA* (Pena et al. 2010).

3.5 Carboxysome

Carboxysomes are intracellular polyhedral proteinaceous compartments (80–140 nm in diameter) inside photosynthetic bacteria machinery, part of the CCM in all cyanobacteria and chemoautotrophs. The polyhedral compartments are tightly packed with RuBisCO enzymes, along with enzyme CA at carboxysomal membrane which helps in fast conversion of HCO_3^- to CO_2. It is believed that these compartments concentrate CO_2 to overcome the catalytic inefficiency of RuBisCO in carbon fixation and allows faster and more efficient CO_2 fixation with less leakage than cytoplasm. It was discovered more than 50 years ago in the cyanobacterium *Phormidium uncinatum* (Drews and Niklowitz 1957). However, first purified form was obtained from *Halothiobacillus neapolitanus* (Shively et al. 1973) and became as a characteristic feature of all cyanobacteria after many genomic studies (Chen et al. 1994). Based partly on the shell protein components and RuBisCO ortholog, carboxysomes have been evident in two evolutionarily distinct groups: α-carboxysomes (prevalent in oceanic cyanobacteria and some chemoautotrophs) and β-carboxysomes (in freshwater/estuarine cyanobacteria) (Tabita et al. 2008). Cyanobacteria with α-carboxysomes occur in environments where dissolved carbon is not limiting (e.g., oligotrophic oceanic waters), whereas cyanobacteria with β-carboxysomes occur in environments where dissolved carbon is limiting (e.g., mats, films, estuaries, and alkaline lakes with higher densities of photosynthetic organisms) (Badger et al. 2002). Although α- and β-carboxysomes shared similarities in external protein shell (Kerfeld et al. 2005), they are different in the interior proteins of the carboxysome compartment. In addition to differing in RuBisCO orthologs, they also diverge in gene organization; genetic components of α-carboxysome organized into an operon, while β-carboxysome gene components are normally more scattered (Kinney et al. 2011). Both evolutionarily distinct forms of carboxysome seem to have evolved twice in parallel as a consequence of fluctuating atmospheric CO_2 levels and evolutionary pressure acting via the poor enzymatic kinetics of RuBisCO. Despite structural distinctions, both carboxysomes are alike in several physiological factors, mostly in the photosynthetic response to light and exterior inorganic carbon (Whitehead et al. 2014). This suggests their different structures today have arisen through a process of convergent evolution and reflects in their distinct evolutionarily strategies to the same major functions: encapsulation of phyletically distinct forms of RuBisCO enzyme (form-1A and 1B), oxygen exclusion, nonidentical shell protein, and enhance CO_2 concentration and fixation.

3.5.1   Elements of Carboxysome Structure

Structurally, typical interior of carboxysome is densely packed in an ordered fashion with its cellular enzymes; major load RuBisCO and a lower concentration of carbonic anhydrase help to efficiently fix the CO_2. These are enclosed within a semi-permeable, icosahedral or quasi-icosahedral; a few thousand proteins shell (80–140 nm in diameter), provide structural organization and prevent CO_2 leakage (Andersson and Backlund 2008). Carboxysomes assemble in a stepwise fashion (produce one structure at a time), inside to outside, revealing that major load RuBisCO is the key coordinator of compartmentalization (Chen et al. 2013) (see Fig. 3.6).

3.5.2   Carboxysomal Shell Proteins

The major structural component of carboxysomes are a 5–6-nm-thin monolayer outer shell (Kaneko et al. 2006), comprised of a orderly arrays of thousands of small (10–12 kD, ~100 amino acid) globular proteins also called bacterial micro compartment (BMC) protein (Cannon and Shively 1983) (see Fig. 3.7). The crystal

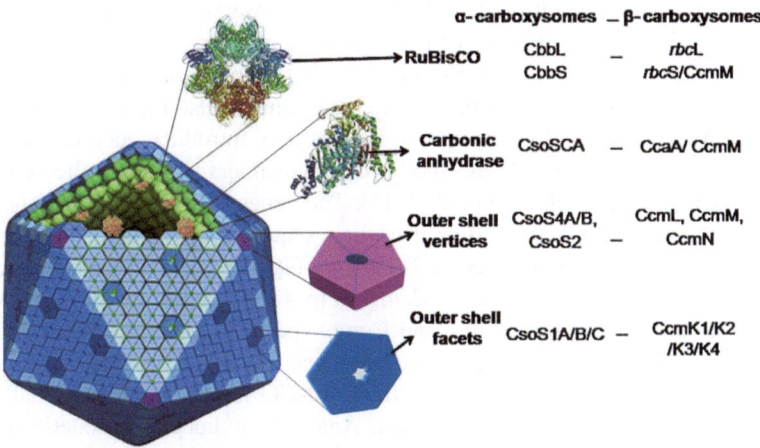

Fig. 3.6 Components of carboxysome structure in cyanobacteria

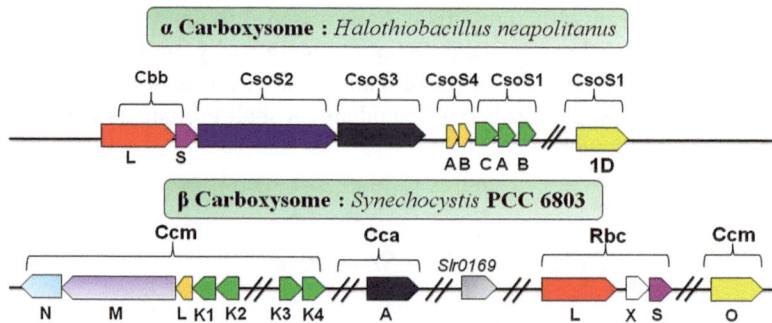

Fig. 3.7 α- and β-carboxysome gene clusters for *Halothiobacillus neapolitanus* and *Synechocystis* **PCC 6803**; *diagonal crossed lines* represent large genomic segments between genes. Homologous proteins are drawn with similar color. The α-carboxysomes, *H. neapolitanus cso* operon contains the genes; form I RuBisCO (*cbbL, cbbS*), pentameric shell-associated proteins (*csoS2*), pentameric shell-associated carbonic anhydrase (*csoS3*), carboxysomal shell proteins (*csoS4A/B*), single-domain BMC hexamers (*csoS1A/B/C*), and permuted trimer (*csoS1D*). The β-carboxysomes, *Synechocystis* PCC 6803 contains the genes RuBisCO (RbcL/X/S), single-domain BMC hexamers (CcmK1/K2/K4), pentameric shell-associated proteins (CcmL), permuted trimer (CcmO), and carbonic anhydrase (CcaA)

forms suggested that individual BMC proteins in carboxysome are tailored to self-assemble side by side to constitute a two-dimensional, molecular layer of cyclic, disc-shaped hexamers, known as the basic building blocks of the carboxysomal shell, tightly packed due to intermolecular hydrogen bonds (Samborska and Kimber 2012). The middle of each BMC hexamer is usually perforated by a narrow well-defined 4-to-7-Å pores within a diameter of about 70 Å, along the six-fold axis of symmetry (Jorda et al. 2013). These apertures may serve as the conduits for diffusion of substrates (bicarbonate and ribulose-1,5-bisphosphate) and products (3-phosphoglycerate) to cross into and out of the carboxysomal shell. The putative aperture is lined by three small residues, including a highly conserved positively charged glycine, which promote the diffusion of the negatively charged substrates and products (Andersson and Backlund 2008). Besides the BMC family, another family of conserved proteins, minor structural pentameric components also present in the shell of all BMCs. Many structural studies proposed that they introduce curvature by bending or folding of flat-layered BMC shell protein hexamers to occupy the vertices in the closed icosahedral shell (Cai et al. 2013). The genes which encode the shell proteins of α and β carboxysome differed in their sequence on chromosome, primary structure, and numbers (Kinney et al. 2011) (see Fig. 3.7). In β carboxysome, these proteins are generally designated as CcmK 1 to 4 (hexameric components with a single BMC domain copy), CcmL, and CcmO (pentameric components with two BMC domains fused head to tail) encoded by *ccm* operon. These gene cluster sets are situated right away upstream from the RuBisCO large (*rbcL*) and small subunit genes (*rbcS*) of cyanobacterial strain. Analysis of knockout genetic mutants from *Synechococcus elongatus* PCC 7942 predicted the function of four proteins (CcmK2-4, CcmO) in the outer shell layer of β-carboxysomes (Rae et al. 2012).

Their results showed that the CcmK2 (forming the bulk facet), CcmO (forming the zipper protein interfacing the edges of carboxysome facets), and CcmL (forming the vertices) genes are minimum structural requirements for the outer shell formation of β-carboxysomes. CcmK2 helps to create an identical coating in which all hexamer protein oriented in similar way, whereas CcmK4 proteins formed a strip of hexamer protein alternating between convex and concave orientations (Kinney et al. 2011). Other proteins, CcmK3, and CcmK4 are necessary to predict the role of preexisting shell. Homologs of these proteins have also been identified in α-carboxysome-containing organisms known as CsoS1 (dual BMC protein), peptide A, and peptide B, encoded by *cso* operon. It is believed that CsoS1 protein may play a role in gated metabolite transport, possibly of the larger RuBisCO metabolites RuBP and 3-phosphoglycerate, across the carboxysome shell (Klein et al. 2009). In contrast to *ccm* operons, *cso* gene clusters of α-carboxysome are situated downstream from the RuBisCO large- and small-subunit genes (cbbL and cbbS, respectively, in chemoautotrophs) (Cannon et al. 2001). The α-carboxysome shells also contain two polypeptides, CsoS2 (80–90 kD) and the CsoS3 (55–65 kD), that have no homology to each other. However, it is believed that in β-carboxysomes, CsoS2, and CsoS3 appear to be replaced by CcmM (55–70 kD) and CcmN (26 kD) (Rae et al. 2013a, b).

3.5.3 Carboxysomal Enzymes

Besides differing in the structural components, α- and β-carboxysomes also vary in encapsulated form of RuBisCO and Carbonic anhydrase.

3.5.3.1 Carboxysomal RuBisCO

The α-carboxysome encapsulates proteobacterial *form-1A* RuBisCO, while the β-carboxysome encapsulates higher-plant *form*-1B RuBisCO (Rae et al. 2013a, b). The genes of the carboxysomal *form-1A* RuBisCO are always part of the *cso* operon, where they are followed by the genes for the α-carboxysomal shell proteins. In contrast, *form*-1B genes of RuBisCO are part of the gene cluster sets encoding the β-carboxysome, only in a few cyanobacterial strains. Several chemoautotrophs (e.g., *Halothiobacillus neapolitanus*) also harbor a gene, *cbbM*, for a chimeric and heterologous species of form-II RuBisCO, and consists of a dimer of large subunits (L_2) (Menon et al. 2008).

3.5.3.2 Carboxysomal Carbonic Anhydrase

In addition to CO_2-fixing enzyme, RuBisCO, all carboxysome also contains an essential β-carbonic anhydrase for the conversion of bicarbonate to CO_2 and may also be play important role in structural integrity of outer shell. It is believed that carbonic anhydrases bind tightly to hexamers pore on the side that faces the inside of the carboxysome. Its catalytic activity might generate a sink that facilitates diffuse out of negative bicarbonate ions through the positively charged pore, across the carboxysome shell. The α-carboxysomes have a distinct CsoSCA (57 kD; encoded by *csoS3*gene) and account for only 2–3 % of carboxysomal protein (Dou et al. 2008). CsoSCA is divergent from β-carbonic anhydrases (Sawaya et al. 2006) that it was initially thought to be the founding member of a new CA class, ε (So et al. 2004). However, in contrast to most β-CAs which arranged as a symmetric dimer with two catalytic sites, CsoSCA is organized as a head-to-tail fusion of two β-CA domains.

The β-carboxysomes have CcmM and CcmN proteins (Ludwig et al. 2000) as an interior organizing structural protein. The CcmM protein (58–94 kD) contains 3–5 repeats of the *rbc*L domain in its C-terminus, while N-terminal domain (~210 residues) is postulated to be an active γ-carbonic anhydrase (approximately 35 % identity) (Alber and Ferry 1994, Long et al. 2010, Kimber 2014). Paradoxically, many β-cyanobacterial strains have lost all CcmM activity and act primarily as a protein complex assembly scaffold. Thus, these domains have been shown to be catalytically active in an organism that lacks the classic carboxysomal carbonic anhydrase (CcaA) ortholog (Pena et al. 2010).

3.6 Functional Carbon-Concentrating Mechanism Model of Cyanobacteria

In the cyanobacteria, CCMs help to tolerate at very low levels of inorganic carbon and efficiently allow photosynthetic CO_2 reduction. Although substantial efforts have been made to explore the CCM, very less information is available for their regulation. Several hypotheses have been suggested for mechanism of sensing the fluctuations in CO_2/HCO_3^- levels, such as photorespiratory metabolite-based mechanism (Huege et al. 2011), direct sensing of the depleted internal C_i pool by HCO_3^--binding sensory protein-like soluble adenylate cyclases (sACs) (Woodger et al. 2005), and redox state changes during electron transport chain of photosynthesis (Rahman et al. 2014). Under high- and low-C_i conditions, different C_i uptake systems of cyanobacteria induced with distinctive uptake flux capacity, and net affinity characteristics (see Fig. 3.8). In *Synechocystis*, cyanobacterial cells grown under high-C_i conditions, where C_i is sufficient, cells typically express only the low-affinity/high-flux C_i transporters such as Na^+-dependent HCO_3^- transporter *bicA*, encoded by *sll0834* (Price et al. 2004), and redox-driven CO_2 uptake system NDH-I_4, encoded by *sll0026*, *sll0027*, and *slr1302* (Shibata et al. 2001, Maeda et al. 2002). As mentioned earlier, NDHI$_4$ complexes are believed to work as CAs, catalyzing the hydration of CO_2. In contrast, higher-affinity/low-flux transporters such as BCT1, encoded by *slr0040-44* (Omata et al. 1999), Na^+-dependent HCO_3^- transporter *SbtA/B*, encoded by *slr1512* and *slr1513* (Shibata et al. 2002, Wang et al.

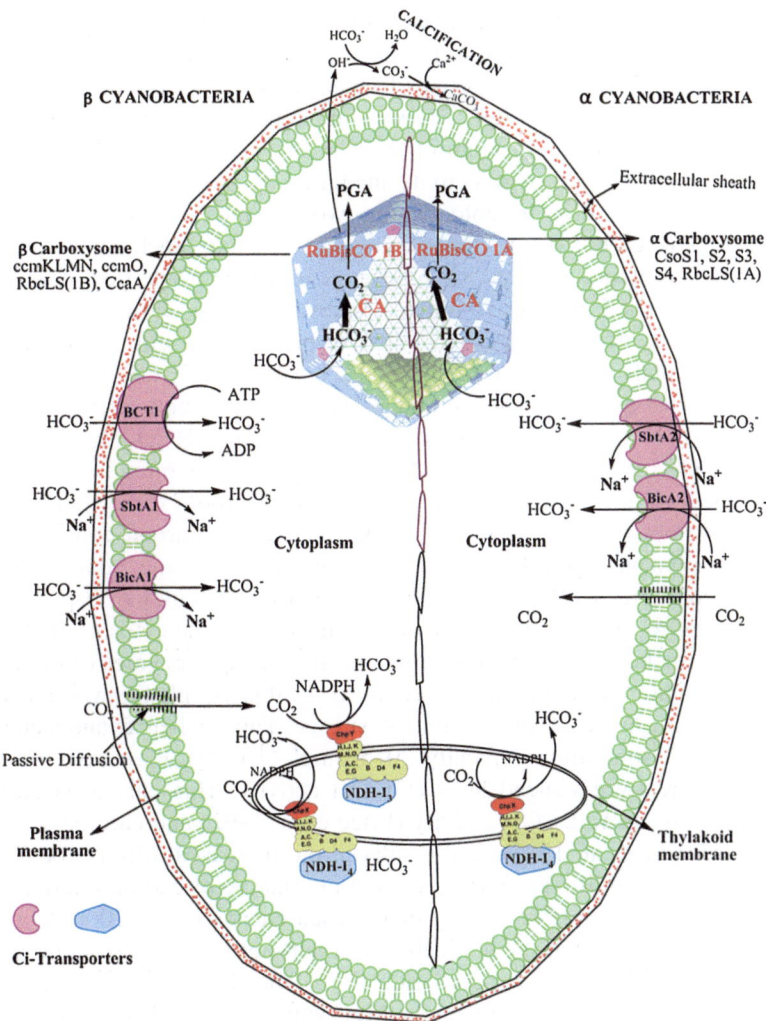

Fig. 3.8 Proposed CCM model of cyanobacteria

2004) and redox-driven NDH-I$_3$, encoded by the genes *sll1732*, *sll1733*, *sll1734*, and *sll1735* (Shibata et al. 2001, Maeda et al. 2002, Klughammer et al. 1999, Prommeenate et al. 2004, Zhang et al. 2004), are additionally expressed upon imposition of low inorganic carbon (low-C$_i$) conditions. After induction of these transporters, C$_i$ is actively pumped into the cell cytosol which corresponds to increase in CCM activity and affinity of C$_i$ transport. Furthermore, HCO$_3^-$ ions spontaneously convert into CO$_2$, through carbonic anhydrases, which rapidly diffused through the membrane. Thus, in low-C$_i$ conditions, inducible CCM systems enable the cyanobacteria to accumulate large internal concentrations of C$_i$ that may be up to 1000-fold higher than the external concentration. Thus, due to

CCM, cyanobacterial cells are able to work with high photosynthetic efficiency, even at low-C_i conditions. This accumulated C_i is required to fix through in vivo photosynthesis. Thus, HCO_3^- traverses further in the carboxysome through small pores in the proteinaceous shell. The key functional feature of carboxysomes is thought to be their ability to control the flux of small molecules into, and out of, the interior by virtue of selectivity mediated by the shell protein pores and enhance the CO_2 fixation by colocalizing two enzymes RuBisCO and carbonic anhydrase. The carboxysome functioning depends on biochemical properties of associated proteins, such as the biochemical/kinetic properties of RuBisCO; conductance of shell proteins to influx of substrate RuBP with respect to efflux of carboxylation product PGA, and influx of bicarbonate with respect to efflux of CO_2, and finally the way through which bicarbonate is transformed to CO_2 inside the carboxysomes. The α- and β-carboxysomes both differ in each of these properties. Inside the icosahedral carboxysome, HCO_3^- is rapidly converted to CO_2 by carboxysomal carbonic anhydrases at rates high enough to saturate the carboxylation reaction of RuBisCO. Carboxysome shells are also postulated to act as a barrier to CO_2 leakage. The five-carbon compound ribulose 1,5-bisphosphate (RuBP) also enters the carboxysome, and sequestered RuBisCO combines RuBP with CO_2 to make two molecules of 3-phophoglyceric acid (3PGA). 3PGA can escape across the shell to the cytosol, mediated by the pores in the hexameric shell proteins. In the cytosol, enzymes of the Calvin–Benson–Bassham cycle regenerate ribulose 1,5-bisphosphate, while diverting the net carbon gained to production of sugars. Thus, when cyanobacteria are subjected to C_i limitation (typically 20–50 mgL^{-1} CO_2), they have the ability to induce a greatly enhanced level of CCM activity. This change is accompanied by increase in RuBisCO activity (Price et al. 1992) and carboxysome shell proteins (McKay et al. 1993), and results in the increased affinity for CO_2 and HCO_3^- uptake. In addition to CCM, another alternative of carbon fixation "calcification" has also been observed by many researchers (Riding 2006; Couradeau et al. 2012). Cyanobacterial calcification is a non-obligate, extracellular process, which occurs in the exopolysacccharide (EPS) sheath or proteinaceous surface layer (S-layer) that surrounds the cells. McConnaughey and Whelan (1997) suggested that calcification may be regarded as a CCM in decreasing photorespiration by the oxygenase activity of RuBisCO, as the protons and CO_2 generated by calcification are used in photosynthesis. The process is stimulated by photosynthetic uptake of C_i and OH^- release from the cell during carbonic anhydrase conversion of HCO_3^- into CO_2; both lead to elevated pH in the sheath. At this raised pH, extracellular HCO_3^- converts into CO_3^{2-}, increasing saturation state with respect to $CaCO_3$ minerals and favoring $CaCO_3$ nucleation in the sheath. Unfortunately, CCM of very few cyanobacterial strains has been studied at the level of biochemistry and physiology, almost exclusively to β-cyanobacteria. Thus, there are noteworthy gaps of CCM information in between α- and β-cyanobacterial strains. Research on cyanobacterial CCMs needs to be extended from freshwater and marine cyanobacterial strains to different varieties of hard water lakes and streams at the present day.

References

Alber BE, Ferry JG (1994) A carbonic anhydrase from the archaeon *Methanosarcina thermophila*. Proc Natl Acad Sci 91(15):6909–6913

Andersson I, Backlund A (2008) Structure and function of RuBisco. Plant Physiol Biochem 46(3):275–291

Badger MR, Hanson D, Price GD (2002) Evolution and diversity of CO_2 concentrating mechanisms in cyanobacteria. Funct Plant Biol 29(3):161–173

Badger MR, Price GD (2003) CO_2 concentrating mechanisms in cyanobacteria: molecular components, their diversity and evolution. J Exp Bot 54(383):609–622

Badger MR, Price GD, Long BM, Woodger FJ (2006) The environmental plasticity and ecological genomics of the cyanobacterial CO_2 concentrating mechanism. J Exp Bot 57(2):249–265

Battchikova N, Aro EM (2007) Cyanobacterial NDH-1 complexes: multiplicity in function and subunit composition. Physiol Plant 131(1):22–32

Battchikova N, Eisenhut M, Aro E-M (2011) Cyanobacterial NDH-1 complexes: novel insights and remaining puzzles. Biochimi Biophys Acta Bioenergetics 1807(8):935–944

Battchikova N, Vainonen JP, Vorontsova N, Keränen M, Carmel D, Aro EM (2010) Dynamic changes in the proteome of *Synechocystis* 6803 in response to CO_2 limitation revealed by quantitative proteomics. J Proteome Res 9(11):5896–5912

Battchikova N, Zhang P, Rudd S, Ogawa T, Aro EM (2005) Identification of NdhL and Ssl1690 (NdhO) in NDH-1L and NDH-1M complexes of *Synechocystis* sp. PCC 6803. J Biol Chem 280(4):2587–2595

Berger S, Ellersiek U, Steinmaller K (1991) Cyanobacteria contain a mitochrondrial complex I-homologous NADH-dehydrogenase. FEBS Lett 286(1):129–132

Bernat G, Appel J, Ogawa T, Ragner M (2011) Distinct roles of multiple NDH-1 complexes in the cyanobacterial electron transport network as revealed by kinetic analysis of P_{700}^+ reduction in various ndh-deficient mutants of *Synechocystis* sp. strain PCC6803. J Bacteriol 193(1):292–295

Cai F, Sutter M, Cameron JC, Stanley DN, Kinney JN, Kerfeld CA (2013) The structure of CcmP, a tandem bacterial microcompartment domain protein from the β-carboxysome, forms a subcompartment within a microcompartment. J Biol Chem 288(22):16055–16063

Cannon GC, Bradburne CE, Aldrich HC, Baker SH, Heinhorst S, Shively JM (2001) Microcompartments in prokaryotes: carboxysomes and related polyhedra. Appl Environ Microbiol 67(12):5351–5361

Cannon GC, Heinhorst S (1804) Kerfeld CA (2010) Carboxysomal carbonic anhydrases: structure and role in microbial CO_2 fixation. Biochimi Biophys Acta Proteins Proteomics 2:382–392

Cannon GC, Shively JM (1983) Characterization of a homogenous preparation of carboxysomes from *Thiobacillus neapolitanus*. Arch Microbiol 134(1):52–59. doi:10.1007/bf00429407

Chen P, Andersson DI, Roth JR (1994) The control region of the pdu/cob regulon in *Salmonella typhimurium*. J Bacteriol 176(17):5474–5482

Chen M, Hiller RG, Howe CJ, Larkum AWD (2005) Unique origin and lateral transfer of prokaryotic chlorophyll-b and chlorophyll-d light-harvesting systems. Mol Biol Evol 22(1):21–28

Chen AH, Robinson-Mosher A, Savage DF, Silver PA, Polka JK (2013) The bacterial carbon-fixing organelle is formed by shell envelopment of preassembled cargo. PLoS ONE 8(9):e76127

Chorus I, Bartram J (1999) Toxic cyanobacteria in water: a guide to their public health consequences, monitoring and management. Spon Press

Colyer CL, Kinkade CS, Viskari PJ, Landers JP (2005) Analysis of cyanobacterial pigments and proteins by electrophoretic and chromatographic methods. Anal Bioanal Chem 382(3):559–569

Cot SSW, So AKC, Espie GS (2008) A multiprotein bicarbonate dehydration complex essential to carboxysome function in cyanobacteria. J Bacteriol 190(3):936–945

Couradeau E, Benzerara K, Gérard E, Moreira D, Bernard S, Brown GE, López-García P (2012) An early-branching microbialite cyanobacterium forms intracellular carbonates. Science 336(6080):459–462

Daley SME, Kappell AD, Carrick MJ, Burnap RL (2012) Regulation of the cyanobacterial CO$_2$-concentrating mechanism involves internal sensing of NADP$^+$ and α-ketogutarate levels by transcription factor CcmR. PLoS ONE 7(7):e41286

Dou Z, Heinhorst S, Williams EB, Murin CD, Shively JM, Cannon GC (2008) CO$_2$ fixation kinetics of *Halothiobacillus neapolitanus* mutant carboxysomes lacking carbonic anhydrase suggest the shell acts as a diffusional barrier for CO$_2$. J Biol Chem 283(16):10377–10384

Drews G, Niklowitz W (1957) Cytology of blue algae. III. Studies on granular inclusions of Hormogonales. Arch Mikrobiol 25(4):333–351

Dudoladova MV, Kupriyanova EV, Markelova AG, Sinetova MP, Allakhverdiev SI, Pronina NA (2007) The thylakoid carbonic anhydrase associated with photosystem II is the component of inorganic carbon accumulating system in cells of halo-and alkaliphilic cyanobacterium *Rhabdoderma lineare*. Biochim Biophys Acta Bioenergetics 1767(6):616–623

Fukuzawa H, Suzuki E, Komukai Y, Miyachi S (1992) A gene homologous to chloroplast carbonic anhydrase (icfA) is essential to photosynthetic carbon dioxide fixation by *Synechococcus* PCC7942. Proc Natl Acad Sci 89(10):4437–4441

Fukuzawa T, Mendez EE, Hong JM (1990) Phase transition of an exciton system in GaAs coupled quantum wells. Phys Rev Lett 64(25):3066

Golecki JR (1979) Ultrastructure of cell wall and thylakoid membranes of the thermophilic cyanobacterium Synechococcus lividus under the influence of temperature shifts. Arch Microbiol 120(2):125–133

Gutu A, Nesbit AD, Alverson AJ, Palmer JD, Kehoe DM (2013) Unique role for translation initiation factor 3 in the light color regulation of photosynthetic gene expression. Proc Natl Acad Sci 110(40):16253–16258

Herranen M, Battchikova N, Zhang P, Graf A, SirpiÃ S, Paakkarinen V, Aro E-M (2004) Towards functional proteomics of membrane protein complexes in *Synechocystis* sp. PCC 6803. Plant Physiol 134(1):470–481

Herrero A, Flores FG (2008) The cyanobacteria: molecular biology, genomics, and evolution. Caister Academic Press, Norfolk

Higgins CF (2001) ABC transporters: physiology, structure and mechanism: An overview. Res Microbiol 152(3):205–210

Hoiczyk E, Baumeister W (1995) Envelope structure of four gliding filamentous cyanobacteria. J Bacteriol 177(9):2387–2395

Hoiczyk E, Hansel A (2000) Cyanobacterial cell walls: news from an unusual prokaryotic envelope. J Bacteriol 182(5):1191–1199

Huege J, Goetze J, Schwarz D, Bauwe H, Hagemann M, Kopka J (2011) Modulation of the major paths of carbon in photorespiratory mutants of *Synechocystis*. PLoS One 6(1):e16278

Jorda J, Lopez D, Wheatley NM, Yeates TO (2013) Using comparative genomics to uncover new kinds of protein-based metabolic organelles in bacteria. Protein Sci 22(2):179–195

Kaneko Y, Danev R, Nagayama K, Nakamoto H (2006) Intact carboxysomes in a cyanobacterial cell visualized by Hilbert differential contrast transmission electron microscopy. J Bacteriol 188(2):805–808

Kerfeld CA, Sawaya MR, Tanaka S, Nguyen CV, Phillips M, Beeby M, Yeates TO (2005) Protein structures forming the shell of primitive bacterial organelles. Science 309(5736):936–938

Kimber MS (2014) Carboxysomal carbonic anhydrases. Carbonic anhydrase: mechanism, regulation, links to disease, and industrial applications. Springer, Berlin, pp 89–103

Kinney JN, Axen SD, Kerfeld CA (2011) Comparative analysis of carboxysome shell proteins. Photosynth Res 109(1–3):21–32

Klein MG, Zwart P, Bagby SC, Cai F, Chisholm SW, Heinhorst S, Cannon GC, Kerfeld CA (2009) Identification and structural analysis of a novel carboxysome shell protein with implications for metabolite transport. J Mol Biol 392(2):319–333

Klughammer B, Sültemeyer D, Badger MR, Price GD (1999) The involvement of NAD (P) H dehydrogenase subunits, NdhD$_3$ and NdhF$_3$, in high-affinity CO$_2$ uptake in *Synechococcus* sp. PCC7002 gives evidence for multiple NDH-1 complexes with specific roles in cyanobacteria. Mol Microbiol 32(6):1305–1315

Ko SBH, Zeng W, Dorwart MR, Luo X, Kim KH, Millen L, Goto H, Naruse S, Soyombo A, Thomas PJ (2004) Gating of CFTR by the STAS domain of SLC26 transporters. Nat Cell Biol 6(4):343–350

Kupriyanova E, Villarejo A, Markelova A, Gerasimenko L, Zavarzin G, Samuelsson Gr, Los DA, Pronina N (2007) Extracellular carbonic anhydrases of the stromatolite-forming cyanobacterium *Microcoleus chthonoplastes*. Microbiology 153(4):1149–1156

Kupriyanova EV, Sinetova MA, Markelova AG, Allakhverdiev SI, Los DA, Pronina NA (2011) Extracellular β–class carbonic anhydrase of the alkaliphilic cyanobacterium *Microcoleus chthonoplastes*. J Photochem Photobiol B 103(1):78–86

Lee RE (2008) Phycology. Cambridge University Press, Cambridge

Long BM, Tucker L, Badger MR, Price GD (2010) Functional cyanobacterial β-carboxysomes have an absolute requirement for both long and short forms of the CcmM protein. Plant Physiol 153(1):285–293

Ludwig M, Suitemeyer D, Price GD (2000) Isolation of ccmKLMN genes from the marine cyanobacterium, *Synechococcus* sp. PCC7002 (Cyanophyceae), and evidence that CcmM is essential for carboxysome assembly. J Phycol 36(6):1109–1119

Maeda S, Badger MR, Price GD (2002) Novel gene products associated with NdhD$_3$/D$_4$-containing NDH-1 complexes are involved in photosynthetic CO_2 hydration in the cyanobacterium, *Synechococcus* sp. PCC7942. Mol Microbiol 43(2):425–435

Maeda S, Price GD, Badger MR, Enomoto C, Omata T (2000) Bicarbonate binding activity of the CmpA protein of the cyanobacterium *Synechococcus* sp. strain PCC 7942 involved in active transport of bicarbonate. J Biol Chem 275(27):20551–20555

McConnaughey TA, Whelan JF (1997) Calcification generates protons for nutrient and bicarbonate uptake. Earth Sci Rev 42(1):95–117

McKay RML, Gibbs SP, Espie GS (1993) Effect of dissolved inorganic carbon on the expression of carboxysomes, localization of RuBisCO and the mode of inorganic carbon transport in cells of the *cyanobacterium synechococcus* UTEX 625. Arch Microbial 159(1):21–29

Menon BB, Dou Z, Heinhorst S, Shively JM, Cannon GC (2008) *Halothiobacillus neapolitanus* carboxysomes sequester heterologous and chimeric RuBisCO species. PLoS ONE 3(10):e3570

Nordlund TM (2011) Quantitative understanding of biosystems: an introduction to biophysics. Taylor & Francis, London

Ogawa T, Kaplan A (2003) Inorganic carbon acquisition systems in cyanobacteria. Photosynth Res 77(2–3):105–115

Ogawa T, Marco E, Orus MI (1994) A gene (*ccmA*) required for carboxysome formation in the cyanobacterium *Synechocystis* sp. strain PCC6803. J Bacteriol 176(8):2374–2378

Ohkawa H, Price GD, Badger MR, Ogawa T (2000) Mutation of ndh genes leads to inhibition of CO_2 uptake rather than HCO$_3$ uptake in *Synechocystis* sp. strain PCC 6803. J Bacteriol 182(9):2591–2596

Omata T, Ogawa T (1986) Biosynthesis of a 42-kD polypeptide in the cytoplasmic membrane of the cyanobacterium *Anacystis nidulans* strain R$_2$ during adaptation to low CO_2 concentration. Plant Physiol 80(2):525–530

Omata T, Price GD, Badger MR, Okamura M, Gohta S, Ogawa T (1999) Identification of an ATP-binding cassette transporter involved in bicarbonate uptake in the cyanobacterium *Synechococcus* sp. strain PCC 7942. Proc Natl Acad Sci 96(23):13571–13576

Omata T, Takahashi Y, Yamaguchi O, Nishimura T (2002) Structure, function and regulation of the cyanobacterial high-affinity bicarbonate transporter, BCT1. Funct Plant Biol 29(3):151–159

Palenik B, Brahamsha B, Larimer FW, Land M, Hauser L, Chain P, Lamerdin J, Regala W, Allen EE, McCarren J (2003) The genome of a motile marine *Synechococcus*. Nature 424(6952):1037–1042

Pena KL, Castel SE, de Araujo C, Espie GS, Kimber MS (2010) Structural basis of the oxidative activation of the carboxysomal β-carbonic anhydrase, CcmM. Proc Natl Acad Sci 107(6):2455–2460

Price GD (2011) Inorganic carbon transporters of the cyanobacterial CO_2 concentrating mechanism. Photosynth Res 109(1–3):47–57

Price GD, Badger MR, Woodger FJ, Long BM (2008) Advances in understanding the cyanobacterial CO_2-concentrating-mechanism (CCM): functional components, Ci transporters, diversity, genetic regulation and prospects for engineering into plants. J Exp Bot 59(7):1441–1461

Price GD, Coleman JR, Badger MR (1992) Association of carbonic anhydrase activity with carboxysomes isolated from the cyanobacterium *Synechococcus* PCC7942. Plant Physiol 100(2):784–793

Price GD, Maeda SI, Omata T, Badger MR (2002) Modes of active inorganic carbon uptake in the cyanobacterium, *Synechococcus* sp. PCC7942. Funct Plant Biol 29(3):131–149

Price GD, Pengelly JJL, Forster B, Du J, Whitney SM, von Caemmerer S, Badger MR, Howitt SM, Evans JR (2013) The cyanobacterial CCM as a source of genes for improving photosynthetic CO_2 fixation in crop species. J Exp Bot 64(3):753–768

Price GD, Sültemeyer D, Klughammer B, Ludwig M, Badger MR (1998) The functioning of the CO_2 concentrating mechanism in several cyanobacterial strains: a review of general physiological characteristics, genes, proteins, and recent advances. Can J Bot 76(6):973–1002

Price GD, Woodger FJ, Badger MR, Howitt SM, Tucker L (2004) Identification of a SulP-type bicarbonate transporter in marine cyanobacteria. Proc Natl Acad Sci 101(52):18228–18233

Prommeenate P, Lennon AM, Markert C, Hippler M, Nixon PJ (2004) Subunit Composition of NDH-1 Complexes of *Synechocystis* sp. PCC 6803 identification of two new ndh gene products with nuclear-encoded homologues in the chloroplast Ndh complex. J Biol Chem 279(27):28165–28173

Rae BD, Long BM, Badger MR, Price GD (2012) Structural determinants of the outer shell of β-carboxysomes in Synechococcus elongatus PCC 7942: roles for Ccm K2, K3–K4, CcmO, and CcmL. PLoS ONE 7(8):e43871

Rae BD, Long BM, Badger MR, Price GD (2013a) Functions, compositions, and evolution of the two types of carboxysomes: polyhedral microcompartments that facilitate CO_2 fixation in cyanobacteria and some proteobacteria. Microbiol Mol Biol Rev 77(3):357–379

Rae BD, Long BM, Whitehead LF, Forster B, Badger MR, Price GD (2013b) Cyanobacterial carboxysomes: microcompartments that facilitate CO_2 fixation. J Mol Microbiol Biotechnol 23(4–5):300–307

Rahman MA, Sinha S, Sachan S, Kumar G, Singh SK, Sundaram S (2014) Analysis of proteins involved in the production of MAA's in two *Cyanobacteria synechocystis* PCC 6803 and *Anabaena cylindrica*. Bioinformation 10(7):449–453

Rasmussen B, Fletcher IR, Brocks JJ, Kilburn MR (2008) Reassessing the first appearance of eukaryotes and cyanobacteria. Nature 455(7216):1101–1104

Riding R (2006) Cyanobacterial calcification, carbon dioxide concentrating mechanisms, and Proterozoic–Cambrian changes in atmospheric composition. Geobiology 4(4):299–316

Rocap G, Distel DL, Waterbury JB, Chisholm SW (2002) Resolution of *Prochlorococcus* and *Synechococcus* ecotypes by using 16S-23S ribosomal DNA internal transcribed spacer sequences. Appl Environ Microbiol 68(3):1180–1191

Samborska B, Kimber MS (2012) A dodecameric CcmK2 structure suggests β-carboxysomal shell facets have a double-layered organization. Structure 20(8):1353–1362

Sandrini G, Matthijs HCP, Verspagen JMH, Muyzer G, Huisman J (2014) Genetic diversity of inorganic carbon uptake systems causes variation in CO_2 response of the cyanobacterium *Microcystis*. ISME J 8:589–600. doi:10.1038/ismej.2013.179

Sawaya MR, Cannon GC, Heinhorst S, Tanaka S, Williams EB, Yeates TO et al (2006) The structure of β-carbonic anhydrase from the carboxysomal shell reveals a distinct subclass with one active site for the price of two. J Biol Chem 281(11):7546–7555

Seckbach J, Oren A, Stan-Lotter H (2013) Polyextremophiles: life under multiple forms of stress. Springer, Berlin

Shelden MC, Howitt SM, Price GD (2010) Membrane topology of the cyanobacterial bicarbonate transporter, BicA, a member of the SulP (SLC26A) family. Mol Membr Biol 27(1):12–22

Shibata M, Katoh H, Sonoda M, Ohkawa H, Shimoyama M, Fukuzawa H, Kaplan A, Ogawa T (2002) Genes essential to sodium-dependent bicarbonate transport in cyanobacteria: function and phylogenetic analysis. J Biol Chem 277(21):18658–18664

Shibata M, Ohkawa H, Kaneko T, Fukuzawa H, Tabata S, Kaplan A, Ogawa T (2001) Distinct constitutive and low-CO_2-induced CO_2 uptake systems in cyanobacteria: genes involved and their phylogenetic relationship with homologous genes in other organisms. Proc Natl Acad Sci 98(20):11789–11794

Shively JM, Ball F, Brown DH, Saunders RE (1973) Functional organelles in prokaryotes: polyhedral inclusions (carboxysomes) of *Thiobacillus neapolitanus*. Science 182(4112):584–586

So AKC, Espie GS (1998) Cloning, characterization and expression of carbonic anhydrase from the cyanobacterium *Synechocystis* PCC6803. Plant Mol Biol 37(2):205–215

So AKC, Espie GS, Williams EB, Shively JM, Heinhorst S, Cannon GC (2004) A novel evolutionary lineage of carbonic anhydrase (ε class) is a component of the carboxysome shell. J Bacteriol 186(3):623–630

So AKC, Van Spall HGC, Coleman JR, Espie GS (1998) Catalytic exchange of 18O from 13C18O-labelled CO_2 by wild-type cells and *ecaA*, *ecaB*, and *ccaA* mutants of the cyanobacteria *Synechococcus* PCC7942 and *Synechocystis* PCC6803. Can J Bot 76(6):1153–1160

Soltes-Rak E, Kushner DJ, Williams DD, Coleman JR (1993) Effect of promoter modification on mosquitocidal cryIVB gene expression in *Synechococcus* sp. strain PCC 7942. Appl Environ Microbiol 59(8):2404–2410

Steinle A, Oppermann-Sanio FB, Reichelt R, Steinbachel A (2008) Synthesis and accumulation of cyanophycin in transgenic strains of *Saccharomyces cerevisiae*. Appl Environ Microbiol 74(11):3410–3418

Tabita FR, Satagopan S, Hanson TE, Kreel NE, Scott SS (2008) Distinct form I, II, III, and IV RuBisco proteins from the three kingdoms of life provide clues about RuBisco evolution and structure/function relationships. J Exp Bot 59(7):1515–1524

Wang H-L, Postier BL, Burnap RL (2004) Alterations in global patterns of gene expression in *Synechocystis* sp. PCC 6803 in response to inorganic carbon limitation and the inactivation of ndhR, a LysR family regulator. J Biol Chem 279(7):5739–5751

Wang J, Chen C (2009) Biosorbents for heavy metals removal and their future. Biotechnol Adv 27(2):195–226

Whitehead AN (2013) The concept of nature. Courier Dover Publications

Whitehead L, Long B, Price D, Badger MR (2014) Comparing the in vivo function of α- and β-carboxysomes in two model cyanobacteria. Plant Physiol 114.237941

Woodger FJ, Badger MR, Price GD (2005) Sensing of inorganic carbon limitation in Synechococcus PCC7942 is correlated with the size of the internal inorganic carbon pool and involves oxygen. Plant Physiol 139(4):1959–1969

Xu M, Bernat Gb, Singh A, Mi H, Ragner M, Pakrasi HB, Ogawa T (2008) Properties of mutants of *Synechocystis* sp strain PCC 6803 lacking inorganic carbon sequestration systems. Plant Cell Physiol 49(11):1672–1677

Yu J-W, Price GD, Song L, Badger MR (1992) Isolation of a putative carboxysomal carbonic anhydrase gene from the cyanobacterium *Synechococcus* PCC7942. Plant Physiol 100(2):794–800

Zarzycki J, Axen SD, Kinney JN, Kerfeld CA (2013) Cyanobacterial-based approaches to improving photosynthesis in plants. J Exp Bot 64(3):787–798

Zhang P, Battchikova N, Jansen T, Appel J, Ogawa T, Aro E-M (2004) Expression and functional roles of the two distinct NDH-1 complexes and the carbon acquisition complex NdhD$_3$/Ndh F$_3$/CupA/Sll1735 in *Synechocystis* sp PCC 6803. Plant Cell Online 16(12):3326–3340

Zhang P, Battchikova N, Paakkarinen V, Katoh H, Iwai M, Ikeuchi M, Pakrasi H, Ogawa T, Aro E (2005) Isolation, subunit composition and interaction of the NDH-1 complexes from *Thermosynechococcus elongatus* BP-1. Biochem J 390:513–520

Chapter 4
Carbon-Concentrating Mechanism of Microalgae

4.1 Introduction

Microalgae (or microphytes) are diverse group of microscopic prokaryotic and eukaryotic photosynthetic microorganisms, typically existed individually or groups in freshwater and marine systems (Andersen 1992). However, microalgae must obtain CO_2 from C_i sources (CO_2 and HCO_3^-), dissolved in the aqueous environments. Many microalgal species have developed an active transport system, CO_2-concentrating mechanism (CCM), to enhance their photosynthetic performance in inadequate CO_2 supply of aquatic ecosystem (Spalding 2008). Since the discovery of the microalgal CCMs, significant effort has been devoted in understanding the molecular mechanism of C_i uptake systems. Although CCMs in eukaryotic microalgae (CO_2 accumulation factor ~180) are not as efficient as cyanobacteria (CO_2 accumulation factor ~800–900) means not effective at raising the internal CO_2 levels, but for most examined algal genera values are considerably lower than this (Beer et al. 2014). In recent years, understanding of C_i transport in the eukaryotic microalgae has been gained largely from *Chlamydomonas reinhardtii*, a unicellular green alga. This alga has been used as a new molecular tool to study the eukaryotic CCM for decades because of its known genome sequence. This is the only recognized eukaryote in which all three genomes (nuclear, mitochondrial, and chloroplast genomes) can be transformed (Dent et al. 2001). In this chapter, we reviewed the most recent and pertinent data concerning the carbon CCM in *C. reinhardtii*.

4.2 Morphological Features of Microalgae

Depending on the species, microalgae sizes can range from a few micrometers (μm) to a few hundreds of micrometers (mm). However, all eukaryotic algae share some basic features such as: a thin, rigid cellulosic cell wall,

© The Author(s) 2014
S.K. Singh et al., *Photosynthetic Microorganisms*, SpringerBriefs in Materials,
DOI 10.1007/978-3-319-09123-5_4

typical eukaryotic nucleus with pores, chromatin, nucleolus, karyolymph, lamellar, discoid or tubular cristae shaped mitochondria, and chloroplasts with enclosed thylakoid as a place of photosynthesis (See Fig. 4.1). Some species may also have one or more flagella for cell motility and a dense, proteinaceous area called pyrenoid, usually associated with synthesis and storage of the starch (Mukherjee →2013). Model microalga *C. reinhardtii* is a single-celled, biflagellate green alga with a cell wall about 10 μm in diameter, surrounding the cytoplasm and a central nucleus (Fan et al. 2012). Cell wall is mainly composed of hydroxyproline-rich glycoprotein and non-cellulosic polysaccharides, as a substitute for cellulose. Both flagella originate from a basal granule located in anterior papillate or non-papillate region of cytoplasm and shows typical 9 + 2 arrangement of the component fibrils. At near the bases of flagella, few contractile vacuoles also occur, which plays an important part in osmoregulation (Hoog et al. 2014). It has a prominent cup- or bowl-shaped chloroplast (Coragliotti et al. 2011), a large pyrenoid (Meyer et al. 2012), and an "eyespot" that senses light (Trippens et al. 2012). The chloroplast contains bands made of a different number of thylakoids that are not ordered into grana-like arrangements (Coragliotti et al. 2011). The central nucleus is enclosed in chloroplast, contains single large pyrenoid at the posterior end of the chloroplast (Hoek 1995). Pyrenoids are RuBisCO-rich microcompartments that act as CO_2 fixation center in microalgae (Meyer et al. 2012), analogous role to carboxysomes of cyanobacteria. It also helps to formation and deposition of starch. Eyespot present in the anterior portion of the chloroplast (Trippens et al. 2012), composed of two, three, more or less parallel rows of linearly arranged fat droplets. Other organelles like Golgi stacks and ER have also been visualized inside the cell.

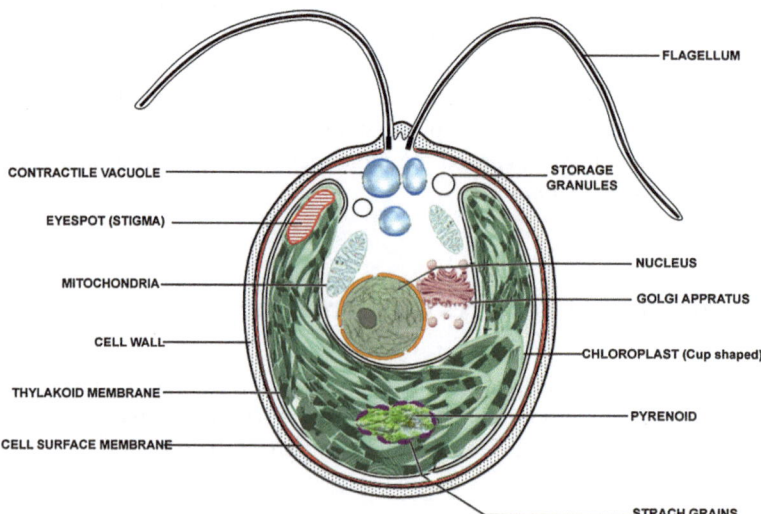

Fig. 4.1 Schematic diagram of a *Chlamydomonas reinhardtii* cell

4.3 Eukaryotic Microalgae C_i Uptake Systems

Among the most critical elements of eukaryotic microalgae's CCM are C_i uptake systems that are responsible for internal C_i accumulation to a stage of many-fold higher than the external medium. In contrast to five types of C_i uptake systems of cyanobacteria, unicellular green alga *C. reinhardtii* have multiple C_i transport systems to acclimate C_i in variable CO_2-limiting conditions. The major mode of C_i uptake into *C. reinhardtii* cells through direct uptake of Ci across the plasma membrane and chloroplast envelope via an active transporter or passively diffusion of CO_2. The movement of CO_2 in and out of the cell is believed to be largely through passive diffusion while the uptake of HCO_3^- ions occurs via an active transporter-mediated process (Spalding 2008).

Many researchers have elucidated expression of several candidate proteins, responsible for C_i uptake, e.g., *LCI1* (Burow et al. 1996), *LCIA* (Miura et al. 2004), *LCIB*, *LCIC*, and *LCID* (Miura et al. 2004; Wang and Spalding 2006), and *HLA3* (Im and Grossman 2002), under limiting-CO_2 growth conditions (see Table 4.1). Under limiting-CO_2 conditions and high external pH levels, candidate genes of CCM are induced and favor the formation of charged HCO_3^- ions. The excitation energy of photosynthesis is used productively for both—C_i uptake against a concentration gradient and CO_2 fixation. Although, numerous putative inorganic transporters localized at the plasma and chloroplast membranes have been studied, candidate transporter proteins for thylakoid membrane are yet to be recognized. Most of the studies reported that C_i transporters of *C. reinhardtii* are expressed under the control of CIA5 gene on chromosome 4 (Xiang et al. 2001; Wang et al. 2005).

4.3.1 Plasma Membrane C_i Transporters

A couple of putative HCO_3^- transporter candidate have been identified in plasma membrane of *Chlamydomonas* transporters: HLA3 and LCI1 (Duanmu et al. 2009a; Ohnishi et al. 2010). The expression of corresponding genes for both transporters has shown increase in response to limited CO_2 conditions. Two Rhesus-like proteins called RHP1 and RHP2 with putative roles as CO_2 channels (Soupene et al. 2004; Kustu and Inwood 2006) are also localized to the plasma membrane (Yoshihara et al. 2008).

4.3.1.1 High Light-Activated 3 (HLA3)

HLA3 (MW 119.7kD) *was* first identified as a high light-induced gene, fell into HLA (1–5) groups (Im and Grossman 2002), whose expression is activated by both high light as well as L-CO_2 conditions. Earlier, it was suggested to be a chloroplast-targeted protein (Im and Grossman 2002). However, deduced HLA3 sequence is predicted by both Target-P (http://www.cbs.dtu.dk/services/TargetP/) and iPSORT (http://hc.ims.u-tokyo.ac.jp/iPSORT/) to be targeted to secretory pathway and thus

Table 4.1 Candidate proteins for C_i uptake systems in eukaryotic microalgae

Protein location	Gene	Comments	References
Chloroplast inner envelope	*Ccp1* and *Ccp2*	L-CO_2 induced, mitochondrial carrier superfamily, identical to each other	Spalding and Jeffrey (1989)
	RHP1 and RHP2	CO_2 channels, 12 transmembrane domains, induction not clear	Amoroso et al. (1998)
	CemA/ycf10	Plastid-encoded gene related to *pxcA* of cyanobacteria and *CemA* of plant plastids; possible function in H^+ extrusion from stroma	Rolland et al. (1997)
	LCIA	Formate/nitrite family of transporters, L-CO_2 induced, most highly remarkable CO_2-responsive gene of all CCM genes, facilitate bicarbonate ions fluxes in *Xenopus*	Galván et al. (2002), Miura et al. (2004), Marsical et al. (2006)
Chloroplast stroma or pyrenoid surroundings	LCIB, LCIC, LCID, LCIE	Unique gene family of soluble proteins, limited to green microalgae	Wang and Spalding (2006), Duanmu et al. (2009a), Yamano et al. (2010)
Plasma membrane	LCI1	Unique gene, increase C_i affinity and accumulation at high pH, second most inducible gene of *C. reinhardtii* CCM	Burow et al. (1996) Ohnishi et al. (2010), Brueggeman et al. (2012)
	HLA3/ MRP1	ATP-dependent ABC-type protein transporters superfamily, activated in both high light as well as L-CO_2 conditions	Im and Grossman (2002); Badger and Price (2003), Badger et al. (2006)

probably to be situated at plasma membrane (Spalding 2008). Similar to BCT1 transporters of cyanobacteria, they have a domain arrangement with high sequence similarity to ATP-binding cassette (ABC)-type transporters superfamily of the multidrug-resistance-proteins (MRP) subfamily (Im et al. 2003). However, quite the opposite to multimeric BCT1 (four protein subunit complex encoded by *cmpABCD* operon) (Badger and Price 2003; Badger et al. 2006), HLA3 is a single protein transporter. There are many genetic and physiological studies supporting the hypothesis that HLA3 is in *C. reinhardtii* essential for bicarbonate transport across the plasma membrane. Duanmu et al. (2009b) confirmed the involvement of HLA3 in C_i uptake during L-CO_2 conditions. Recently, Brueggeman et al. (2012) observed 22–40 times increase in HLA3 transcripts when *C. reinhardtii* cells were shifted high to L-CO_2 level ($\leq$100 mgL^{-1}). Interestingly, mutants lacking the *CIA5* gene show no induction of HLA3 following a shift to L-CO_2 concentrations (Miura et al. 2004; Wang et al. 2005; Fang et al. 2012). Thus, similar to various CCM genes, it appeared to be under the control of CIA5 gene. However, further biochemical characterization of HLA3/MRP1 is still needed to elucidate the mechanism of this putative plasma membrane HCO_3^- transporter. The HLA3 homologs were also reported in *Volvox carteri*, *Chlorella sp*. NC64A, and *Ostreococcus* RCC809 (Yamano and Fukuzawa 2009).

4.3.1.2   Limiting-CO$_2$-inducible 1

The limiting-CO$_2$-inducible 1 (LCI1) was first identified by Burow et al. (1996), localized on plasma membranes (Ohnishi et al. 2010). Ohnishi et al. (2010) observed increase in C$_i$ affinity and C$_i$ uptake in over-expression strains, in contrast to LCR1 mutant under limiting-CO$_2$ environments. The LCI1 transcripts reported to be increase in C$_i$ affinity and C$_i$ accumulation at high pH, suggesting it may have an essential role as a novel plasma membrane HCO_3^- transporter. There is a 3,000 times increase in LCI1 transcripts when the CCM is induced (Brueggeman et al. 2012), makes this the second most inducible gene of *C. reinhardtii* CCM. However, in contrast to HLA3, it encodes a unique protein of 192 amino acids that appears to be no identifiable homologue or domains to any other proteins in available databases Ohnishi et al. (2010). Similar to other CCM genes, LCI1 is also regulated by "master regulator" gene, CIA5 (Im et al. 2003; Miura et al. 2004) and additionally a myb-transcription factor, LCR1 (downstream of CIA5) (Yoshioka et al. 2004).

4.3.1.3   RHP1 and RHP2

The *C. reinhardtii* also encodes the two Rhesus-like proteins, namely RHP1 and RHP2, which have resemblance to Rh proteins of the human red blood membranes (Soupene et al. 2004; Kustu and Inwood 2006). Both are predicted to have 12 transmembrane domains and localized to the plasma membrane (Yoshihara et al. 2008). The RHP1 gene is relatively expressed high under H-CO$_2$ conditions than L-CO$_2$ conditions, suggesting its probable function in CO$_2$ uptake by the cell, under abundant CO$_2$ conditions. No evidence so far supports a role for RHP1 and/or RHP2 in the facilitated CO$_2$ diffusion in L-CO$_2$ conditions (Wang et al. 2011). The continued low expression of RHP1 and RHP2 under L-CO$_2$ conditions does not indicate their large involvement in CO$_2$ transport for the CCM. Nevertheless, their potential role as a CO$_2$ channel cannot be ruled out completely, as both expressed to a few extent in L-CO$_2$.

4.3.2   Chloroplast C$_i$ Transporter

It is now universally established fact that chloroplasts are crucial in C$_i$ concentrating and active C$_i$ uptake at chloroplast's envelope, demonstrated by isolated intact chloroplasts (Amoroso et al. 1998). Several promising candidate genes and putative transporters have been identified as playing a role in the chloroplast membrane bound C_i transport system, including LCIA (NAR1.2) (Miura et al. 2004, Mariscal et al. 2006) belonging to the formate/nitrite (FNT) family of transporters, unique LCIB protein family LCIC/LCID/LCIE (Miura et al. 2004; Wang and Spalding 2006), *Ccp1/Ccp2* belonging to the mitochondrial carrier protein family (Grossman et al. 2007; Jungnick et al. 2014), and *ycf10* belonging to the chloroplast envelope membrane protein A (*CemA*) family (Rolland et al. 1997).

4.3.2.1 LCIA

The low-carbon-inducible-A (LCIA) protein also annotated as NAR1.2, was first identified by Miura et al. (2004), shows sequence similarity to NAR1.1, a chloroplast formate/nitrite transporter family (Galván et al. 2002). LCIA is predicted to be a low-C_i-induced hexameric transmembrane channel at the chloroplast membrane and believed to facilitate bicarbonate ions fluxes rather than active transport (Wang et al. 2009), across the chloroplast envelope in limiting-CO_2-acclimated cells. There is a 3,300-fold at 60 min and 4,000-fold at 180 min increase in LCIA transcripts when the CCM is induced under limiting-CO_2 conditions (Brueggeman et al. 2012), making this the most highly remarkable CO_2-responsive gene of all CCM genes. Like most of the other CCM genes, LCIA is induced under the control of CIA5.

4.3.2.2 LCIB, LCIC, LCID, and LCIE

The low-carbon-inducible-B (LCIB) protein is a soluble protein that lacks any obvious transmembrane domain, localized either in chloroplasts stroma (Duanmu et al. 2009a) or on the pyrenoid surroundings (Yamano et al. 2010). They appear to be limited to green microalgae with CCMs, show a very limited phylogenetic distribution outside these groups. Due to the absence of recognizable motifs and substantial similarity in amino acid sequence, they become part of a unique gene family of soluble proteins, responsive to limiting CO_2, including three homologous genes LCIC (73 % similarity; 57 % identity), LCID, and LCIE (78 % similarity; 71 % identity) in *Chlamydomonas* (Miura et al. 2004; Wang and Spalding 2006). Yamano et al. (2010) reported that LCIB and LCIC form the most abundant transcripts, a 350-KD hexameric complex, expressed very low in high-CO_2 (H-CO_2) and highly expressed in both low CO_2 (L-CO_2) and very low CO_2 (VL-CO_2). This complex could relocalize from the chloroplast stroma to the pyrenoid surrounding and vice versa, under different CO_2 conditions (Fang et al. 2012). When CO_2 concentration changed from H-CO_2 to L-CO_2, LCIB/LCIC complex relocalized from the chloroplast stroma to the pyrenoid surrounding, while, in H-CO_2 or dark conditions, complex diffused rapidly from around the pyrenoid to the stroma (Yamano et al. 2010). Thus, LCIB/LCIC complex works as a structural barrier to prevent CO_2 leakage and sustain CO_2 levels in the matrix of pyrenoid for carbon fixation. In contrast to LCIB/LCIC complex, LCIB/LCID complex transcripts are almost undetectable by standard northern analysis (Wang and Spalding 2006). Thus, LCIB seems to act as either a regulator or an integral component of a multisubunit plastid complex that helps in C$_i$ transport within the chloroplast or retains CO$_2$ in the chloroplast (Yamano and Fukuzawa 2009). Duanmu et al. (2009b) showed that probably LCIB also involves in epistatically downstream of CAH3, a carbonic anhydrase localized in the thylakoid lumen that dehydrates the HCO$_3^-$ and release CO$_2$ from the lumen (Duanmu et al. 2009a). LCIB protein could help to prevent CO$_2$ leakage through recapture of excess CO$_2$, released due to CAH3. Thus, in the albescence of LCIB, an ineffective sequence would be created among the active Ci uptake and rapid CO$_2$ release by CAH3. Therefore, LCIB's function in C$_i$ transports is still to be verified experimentally. Other two LCID and LCIE proteins' role is also unknown.

4.3.2.3 Ccp1 and Ccp2

The chloroplast carrier proteins 1 and 2 (*Ccp1* and *Ccp2*) were first discovered by Spalding and Jeffrey (1989), localized at the chloroplast envelope (Brueggeman et al. 2012) with 96 % amino acid sequence similarity to mitochondrial carrier proteins (MCP) superfamily. Brueggeman et al. (2012) reported that under induced CCM conditions in *Chlamydomonas*, *Ccp1* transcripts increased to 2,000-fold, which is very high than *Ccp2*'s 120-fold, makes *Ccp1* the third most highly expressed protein of microalgal CCM. However, earlier, RNA$_i$-mediated knock-down of *Ccp1* and *Ccp2* study by Pollock (Pollock et al. 2004) showed decrease in abundance of the mRNAs for *Ccp1* and *Ccp2* and little or no evidence of impaired inorganic uptake. Thus, it may also be possible that C$_i$ transporter functions of *Ccp1* and *Ccp2* might be masked by some other accumulation systems or C$_i$ transporter. Therefore, *Ccp1* and *Ccp2* proteins' role in C$_i$ transport is still unexplored area of microalgal CCM research.

4.3.2.4 CemA/Ycf10

The chloroplast envelope membrane protein A (*CemA*) is an integral membrane proton pump (encoded by *ycf10*) localized in the inner chloroplast envelope, believed to be pump protons out of the stroma during chloroplast Ci uptake (Rolland et al. 1997; Kaplan and Reinhold 1999). Thus, it may essentially require for maintaining the alkaline stromal pH for functional C$_i$ accumulation in chloroplasts as dehydration of bicarbonate releases hydroxyl anions that have to be exported from the chloroplast or balanced by an equimolar influx of protons. *CemA* is homologous to cyanobacterial *PxcA* (*CotA*) protein, a light-induced Na$^+$-dependent proton pump (Sonoda et al. 1998), which also play important role to maintain electrical and pH homeostasis during C$_i$ uptake. Nevertheless, contrasting to *Ccp1*, *Ccp2*, and LCIA, *CemA* is not regulated by CIA5 (Ohnishi et al. 2010).

Table 4.2 Carbonic anhydrase gene families of *Chlamydomonas reinhardtii*

CA family	Gene	Chromosome	M.W. (kD)	Location	Reference
α	CAH1	4	78	Periplasm	Fujiwara et al. (1990)
	CAH2	4	84	Periplasm	Rawat and Moroney (1995)
	CAH3	9	29.5	Thylakoid lumen	Karlsson et al. (1998)
β	CAH4	5	21	Mitochondria	Eriksson et al. (1996)
	CAH5	5	21	Mitochondria	Eriksson et al. (1996)
	CAH6	12	31	Chloroplast stroma	Mitra et al. (2004)
	CAH7	13	35.79	??	Ynalvez et al. (2008)
	CAH8	9	35.79	Plasma membrane/ periplasm	Ynalvez et al. (2008)
	CAH9	5 + scaffold 61	13.06	Cytoplasm	Merchant et al. (2007)
γ	CAG1	9	24.29	Mitochondria	Cardol et al. (2005), Merchant et al. (2007)
	CAG2	6	31.17	Mitochondria	Cardol et al. (2005), Merchant et al. (2007)
	CAG3	12	32.69	Mitochondria	Cardol et al. (2005), Merchant et al. (2007)

4.4 Algal Carbonic Anhydrases

The identity, number, and localization of carbonic anhydrases (CA) in each gene family (α, β, and γ–CAs) vary with the species in eukaryotic photosynthetic microorganisms. The individual cells may contain CAs from all three families or multiple isoforms of a single CA family, e.g., smaller green alga *Ostreococcus lucimarinus* (1 β- and 1 γ-CAs) (Palenik et al. 2007; Peers and Niyogi 2008), fresh water green alga *Chlorella* (4 α-, 3 β-, and 3 γ-CAs), *Arabidopsis* (6 β- and 5 γ-CAs), *C. reinhardtii* (3 α-, 6 β-, and 3 γ-CAs) (Moroney et al. 2011), etc. The Table 4.2 shows the carbonic anhydrase gene families of *C. reinhardtii*.

4.4.1 α-CAs

In the nineties, three of α-CA isoforms have been identified; two CAH1 and CAH2 at periplasmic membrane (Kimpel et al. 1983; Fujiwara et al. 1990; Fukuzawa et al. 1991) and CAH3 at chloroplast thylakoid isoform (Karlson et al. 1995).

4.4.1.1 CAH1

The CAH1 is among the earliest α-CA genes reported for a photosynthetic organism, highly induced in grown cells under L-CO_2 as well as light conditions while completely suppressed under H-CO_2 conditions (Fujiwara et al. 1990). Brueggeman et al. (2012) reported, through RNA-Seq experiments, a high level of expression (approximately 1,000 time increase) of CAH1 in 3 h of L-CO_2 conditions. Earlier, it was believed that they assist entry of CO_2 into the algal cell through periplasmic membrane as CAH1 inhibition has shown a corresponding decrease in C_i affinity of L-CO_2-acclimated *C. reinhardtii* cells (Berry et al. 1976; Badger et al. 1980; Tsuzuki and Miyachi 1989; Moroney et al. 1985). However, contradictory, CAH1 mutant studies of Van and Spalding (1999), not succeeded to demonstrate any quantifiable distinction in C_i affinity between wild- and mutant-type cells. Thus, precise role of CAH1 in the *C. reinhardtii* CCM is still mysterious. CAH1 is up-regulated directly by CCM "master regulator" CCM1 (PMID: 11287669) and their downstream myb-transcription factor LCR1 (PMID: 15155888) (Yoshioka et al. 2004).

4.4.1.2 CAH2

Another periplasmic α-CA isoform, CAH2, is believed to be the consequence of tandem gene duplication event (Grossman et al. 2007), located just 1.4 kb away from CAH1 gene and shows a very close sequence similarity to CAH1. However, unlike CAH1, CAH2 expression is very week under L-CO_2 conditions, induced barely under H-CO_2 conditions. Thus, CAH2 has no role in *C. reinhardtii*'s CCM (Fujiwara et al. 1990; Rawat and Moroney 1995; Brueggeman et al. 2012).

4.4.1.3 CAH3

The third 29.5 kDa CA of *Chlamydomonas* α-CA isoforms, CAH3, is a vital element of CCM, as mutants without CAH3 unable to survive in limiting-CO_2 conditions (Karlsson et al. 1998; Hanson et al. 2003). They localized in lumen of thylakoids traversing the RuBisCO packed-pyrenoid structure (Moroney and Ynalvez 2007) and involved in catalyzing the HCO_3^- ions conversion (brought in from the chloroplast stroma) to CO_2 (Raven 1997; Karlsson et al. 1998; Shutova et al. 2008). The closer proximity of CAH3 to RuBisCO molecule ensures quicker CO_2 delivery and fixation with less leakage, under limiting-CO_2 conditions (Sinetova et al. 2012). In L-CO_2 conditions, *CAH3* expressed twofold more in transcripts abundance than H-CO_2 (Blanco-Rivero et al. 2012; Sinetova et al. 2012). The CAH3 might also be associated with O_2-evolving photosystem II (PS-II) complex (Park et al. 1999). They stabilize the Mn^{2+} cluster of PS-II by supplying the HCO_3^- to acceptor, which facilitates the electron flow between Q_A and Q_B (Shevela et al. 2012). Duanmu et al. (2009a) have also shown another function of CAH3 in an epistatic manner to LCIB protein. However, the exact physiological role of CAH3 in the CCM is yet to be explored. Similar to CAH1, the expression of CAH3 gene is also controlled by CIA5 (Ynalvez et al. 2008).

4.4.2 β-CAs

The six β-CA isoforms, CAH4, CAH5, CAH6, CAH7, CAH8, and CAH9 have been reported in *C. reinhardtii*.

4.4.2.1 CAH4 and CAH5

The first β-CAs to be discovered were two L-CO_2-induced mitochondrial proteins, CAH4 and CAH5. They were isolated from mitochondrial preparations of *Chlamydomonas* (Eriksson et al. 1996), were verified for their subcellular localization in mitochondria (Cardol et al. 2005) The CAH4 and CAH5 proteins (~21 kD) were almost similar, vary by only a single amino acid (Eriksson et al. 1996; Villand et al. 1997) and organized in an inverted repeat (head to head) pattern on chromosome 5, demonstrating a probable tandem gene duplication incident. Both CAs are one of the strongly up-regulated CCM proteins (Villand et al. 1997; Erikkson et al. 1998), showed up-regulation of more than 512-fold in *Chlamydomonas* cells after a switch from H-CO_2 to L-CO_2 (Brueggeman et al. 2012; Fang et al. 2012). In contrast, the transcript levels of both were almost undetectable when cells switched to H-CO_2 conditions (Fujiwara et al. 1990). Furthermore, like CAH1, both CAs once produced in *C. reinhardtii* cells grown under L-CO_2, retained for 4 days at least, after restoration to H-CO_2 conditions (Moroney et al. 2011). This might be explained by peripheral migration of mitochondria under L-CO_2 conditions, which makes them a potential candidate for CO_2 leakage, if allowed to diffuse out of the mitochondria. In limiting-CO_2 conditions, equilibrium among carbon and nitrogen pathways disturbed significantly. Thus, as an overall stress response by *Chlamydomonas* cells, there is an urgent requirement of carbon recycling for nitrogen assimilation. The CAH4 and CAH5 are believed to play a vital role in not only buffering the pH of mitochondrial matrix to neutralize ammonia (NH_3) liberated through photorespiration processes (Raven et al. 2002) but also in retaining and anaplerotic carbon recycling of generated CO_2 (by hydration of CO_2 to HCO_3^-) during photorespiration (Giordano et al. 2003). In the view of mitochondrial migration nearer to plasma membrane of cell under L-CO_2 conditions, CCM proteins CAH4 and CAH5 roles might be related to prevent CO_2 leakage from the cells or trapping any incoming CO_2 and released further in the form of HCO_3^- into the cytosol for uptake by the chloroplast (Giordano et al. 2003; Moroney et al. 2011). Thus, CAH4 and CAH5 are not directly associated with CCM but they facilitate to yield additional CO_2 as photosynthetic substrate for RuBisCO (Giordano et al. 2003).

4.4.2.2 CAH6

The CAH6 is another important constitutively expressed chloroplastic β-CA (Mitra et al. 2004). It plays a direct role in *C. reinhardtii* CCM as its absence has direct physiological effects on L-CO_2-acclimated *C. reinhardtii* cells (Moroney et al. 2011). Like CAH3, CAH6 is expressed in both H-CO_2 and L-CO_2 conditions, but expression increased in L-CO_2 (Ynalvez et al. 2008; Brueggeman et al. 2012; Fang et al. 2012). Immunogold labeling studies showed CAH6 location in the chloroplast stroma of *C. reinhardtii* cells, predominantly distributed near starch sheath that covered the pyrenoid (Mitra et al. 2004). But, localization of CAH6 is still a topic of speculation. The main CCM role of CAH6 in the *C. reinhardtii* appears to be recapturing and hydration of CO_2 as it diffuses into the chloroplast stroma from either cytosol or RuBisCO inside pyrenoid (Moroney and Ynalvez 2007). Thus, CAH6 helps to prevent diffusive loss of CO_2 out of the cell and provide a steep gradient for CO_2 diffusion into the chloroplast stroma (Mitra et al. 2004, 2005).

4.4.2.3 CAH7 and CAH8

The CAH7 and CAH8 are strongly interrelated β-CAs with 63 % resemblance (Mitra et al. 2004; Ynalvez et al. 2008) and long hydrophobic C-terminal extensions with transmembrane helices (Ynalvez et al. 2008). Both have not yet been convincingly localized while through immunogold labeling studies, CAH8 is believed to be present at or near the plasma membrane (Moroney et al. 2011), suggesting facilitate C_i entry into the cell. However, CAH7 and CAH8 are constitutively expressed at moderate amounts in *C. reinhardtii* cell. Thus, their probable physiological roles in facilitation of C_i uptake may be secondary to CAH1. But, this is also yet to be verified experimentally.

4.4.2.4 CAH9

The CAH9 is a recently reported cytosolic β-CA (Moroney et al. 2011), expressed very low in *Chlamydomonas* cells under the L-CO_2 conditions. This suggest that CAH9 does not have any significant role in CCM (Ynalvez et al. 2008; Moroney et al. 2011; Brueggeman et al. 2012; Fang et al. 2012).

4.4.3 γ-CAs

The three genes encoding γ-CAs, CAG1, CAG2, and CAG3 have also been reported in *C. reinhardtii*, expressed constitutively without any significant response to changes in CO_2 levels. CAG3 γ-CA shares 30–40 % amino acid sequence resemblance with γ-CA of cyanobacteria (Price et al. 1993; Brueggeman

2013). But, subcellular location and CA activity in the *C. reinhardtii* CCM are still uncertain. Like higher plants, they are believed to be part of mitochondrial Complex I of electron transport chain at the membrane (Cardol et al. 2004, 2005).

4.5 Pyrenoid

The pyrenoids are specialized RuBisCO-rich microcompartments found in chloroplasts of many microalgae that act as center of CO_2 fixation, analogous role to carboxysomes of cyanobacteria (Meyer et al. 2012). It is believed that 90 % part of this suborganellar is composed of photosynthetic enzyme RuBisCO (Nassoury et al. 2001). Carbonic anhydrase also closely associated with RuBisCO to facilitate the conversion of HCO_3^- to CO_2 and maintaining a CO_2-rich environment around the RuBisCO (Maren 1967).

4.5.1 Structural Organization

The pyrenoids are extremely plastic organizations and the amount of RuBisCO packaging inside associates with the status of CCM induction (Ma et al. 2011). The number, morphology, and ultrastructure of pyrenoid may vary substantially in between algal species, e.g., green alga *C. reinhardtii* (Ramazanov et al. 1994) and unicellular red alga *Porphyridium purpureum* have single noticeable pyrenoid in a single chloroplast (Schornstein and Scott 1982) while diatoms and dinoflagellates may be with multiple pyrenoids (Dodge 1968). Despite these variations, all pyrenoids have some common fundamental characteristics as pyrenoid matrix, composed primarily of biochemically active RuBisCO, which is often traversed by stromal thylakoids (Ma et al. 2011). Unlike carboxysomes, they are membrane less bodies. Usually, they are surrounded by a starch sheath, formed or deposited at the periphery of pyrenoids (Ramazanov et al. 1994). An additional concentric layer of LCIB/LCIC protein complex also forms outside the starch sheath under L-CO_2 conditions (Mukherjee 2013), act as a CO_2-leakage barrier or possible recapture CO_2 escaping out from the pyrenoid, together with the stromal carbonic anhydrase CAH6 (Yamano and Fukuzawa 2009). Mutagenic studies on *Chlamydomonas* have revealed that mutants that lack RuBisCO have been shown to lack a pyrenoid structure (Rawat et al. 1996), especially the small subunit of RuBisCO is important for pyrenoid assembly. However, some of the queries like whether RuBisCO self-assembles into pyrenoids or requires additional chaperones still remain to be answered. Another protein CIA6, encoding a putative SET-domain methyl transferase, is reported to require for organized pyrenoid, despite having normal levels of RuBisCO (Ma et al. 2011). Thus, characterizing the entire protein assortment and biochemical composition of pyrenoid is now an active area of research and yet to be fully elucidated.

4.5.2 Physiological Role of Pyrenoid in CCM

On the basis of many metabolic and mutagenic studies, it appears that the main physiological role of pyrenoid is to maintain an intracellular pool of C_i (mainly in HCO_3^- form) and to provide a CO_2-rich environment around the RuBisCO under L-CO_2 conditions of aqueous environment (Mukherjee 2013). Under L-CO_2 conditions, a number of morphological changes occurred in cells including formation of a starch sheath around the pyrenoid, reorganization of RuBisCO, movement of LCIB/LCIC complex closer to the pyrenoid, etc. Reorganization of RuBisCO results the pyrenoid localization as opposed to chloroplast stroma. The LCIB/LCIC protein complex together with the stromal carbonic anhydrase CAH6 also plays an important role in possible recapturing of CO_2 that is leaking out of the pyrenoid (Yamano and Fukuzawa 2009). The CAH6 catalyzes the interconversion of HCO_3^- into CO_2 and results in a high local CO_2 concentration, which RuBisCO can utilize by fixing the CO_2 before it has a possibility to escape out of the *Chlamydomonas* cell. Thus, pyrenoid enhances the efficacy of photosynthesis in an aqueous environment.

4.6 Functional Carbon-Concentrating Mechanism Model of Microalgae

The eukaryotic microalgae naturally live in CO_2-limited environment, as the CO_2 diffusion rate in water is ten thousand times slower than in air (Evans and Von Caemmerer 1996). Thus, they are dependent on CO_2–CCM for sufficient C_i supply. The CCM is induced by low CO_2 and increased the internal concentration of CO_2 in vicinity of ribulose 1, 5-bisphosphate (RuBP) (Ma et al. 2011). In *C. reinhardtii*, many CCM-related proteins have been identified by comparing their expression levels under L-CO_2 versus H-CO_2 environment (Fang et al. 2012). On the basis of studies and data given in this section, a functional speculative CCM model for *C. reinhardtii* could be proposed and illustrated in Fig. 4.2. The *Chlamydomonas* CCM have many multiple Ci transport systems such as *LCI1, LCIA, LCIB, LCIC, LCID, Ccp1, Ccp2*, and *HLA3*, to acclimate C_i in variable CO_2-limiting conditions. After the CCM induction in limiting-CO_2 conditions, these transporters help to transport HCO_3^- ions actively across the plasma membrane (HLA3 and LCI1) and inner envelope of chloroplast (LCIA, an LCIB-associated transporters and possibly *Ccp1/2*). In contrast, CO_2 uptake across both the plasma membrane and the chloroplast envelope occurs by diffusion through the membrane or CO_2 channels (RHP1 and RHP2). The C_i transport via transporters at the plasma membrane, chloroplast envelope, thylakoid membranes, and carbonic anhydrases in the periplasm, cytoplasm, and chloroplast stroma are believed to be contributing an intracellular pool of C_i (mainly HCO_3^-) accumulation in the stroma. High external pH levels also favor the formation of charged HCO_3^- ions. The accumulated HCO_3^-

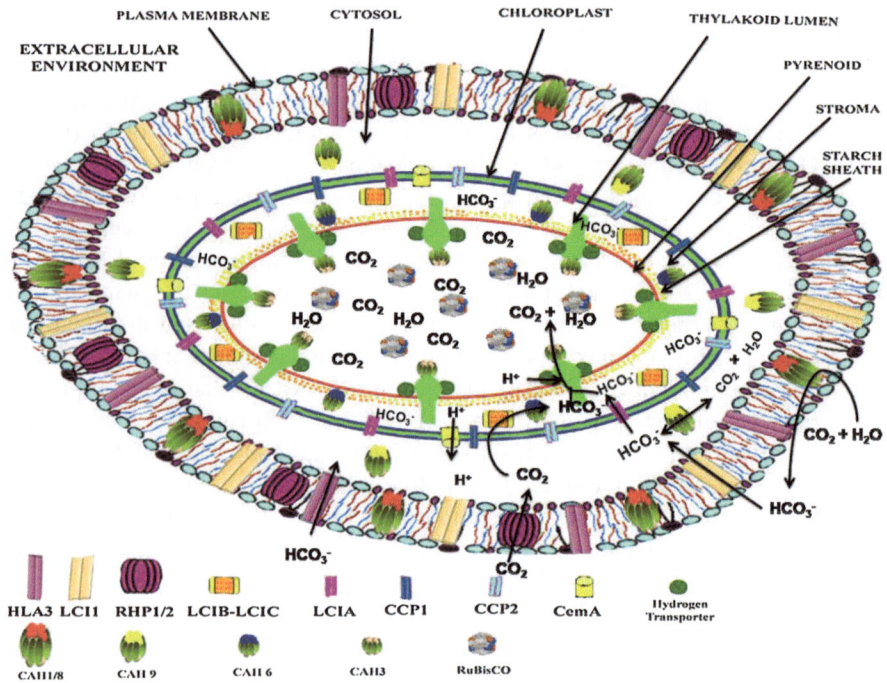

Fig. 4.2 Schematic diagram of CCM model for *C. reinhardti*

thought to be pumped into the acidic lumen of transpyrenoidal thylakoids, where CAH3 converts HCO_3^- to CO_2. This converted CO_2 serves as the substrate source for RuBisCO, located mainly in the pyrenoid. Most of the C_i transporters and carbonic anhydrases expressed in CCM are regulated by a nuclear transcriptional master regulator, CIA5 gene on chromosome 4 (Miura et al. 2004; Brueggeman et al. 2012; Fang et al. 2012). It is believed that CIA5 affects over 1,000 CO_2-responsive genes and also environment for the degree of packing of RuBisCO inside the pyrenoid. However, CIA5 expression is independent of CO_2 concentration. They are incapable to bind DNA strand directly, required additional activation systems to express, such as LCR1, a nuclear transcription factor. LCR1 activates CAH1 and LCI1 (Yoshioka et al. 2004). The CIA5 seems to be conserved only in a green algae class Trebouxiophyceae (*Chlorella* and *Coccomyxa*) of Chlorophyta division, while in *Chlamydomonas* LCR1 is an outlier with CIA5.

References

Amoroso G, Saltemeyer D, Thyssen C, Fock HP (1998) Uptake of HCO_3^- and CO_2 in cells and chloroplasts from the Microalgae *Chlamydomonas reinhardtii* and Dunaliella tertiolecta. Plant Physiol, 116 (1):193–201

Andersen RA (1992) Diversity of eukaryotic algae. Biodivers Conserv 1(4):267–292

Badger MR, Kaplan A, Berry JA (1980) internal inorganic carbon pool of *Chlamydomonas reinhardtii* Evidence for a carbon dioxide-concentrating mechanism. Plant Physiol 66(3):407–413

Badger MR, Price GD (2003) CO_2 concentrating mechanisms in cyanobacteria: molecular components, their diversity and evolution. J Exp Bot 54(383):609–622

Badger MR, Price GD, Long BM, Woodger FJ (2006) The environmental plasticity and ecological genomics of the cyanobacterial CO_2 concentrating mechanism. J Exp Bot 57(2):249–265

Beer S, Bjark M, Beardall J (2014) Photosynthesis in the Marine Environment. Wiley, New Jersey

Berry J, Boynton J, Kaplan A, Badger M (1976) Growth and photosynthesis of *Chlamydomonas reinhardtii* as a function of CO_2 concentration. Annual report

Blanco-Rivero A, Shutova T, Romaịn MJ, Villarejo A, Martinez F (2012) Phosphorylation controls the localization and activation of the lumenal carbonic anhydrase in *Chlamydomonas reinhardtii*. PloS one 7(11):e49063

Brueggeman AJ (2013) Transcriptomic Analyses of the CO_2-Concentrating Mechanisms and Development of Molecular Tools for *Chlamydomonas reinhardtii*

Brueggeman AJ, Gangadharaiah DS, Cserhati MF, Casero D, Weeks DP, Ladunga I (2012) Activation of the carbon concentrating mechanism by CO_2 deprivation coincides with massive transcriptional restructuring in *Chlamydomonas reinhardtii*. The Plant Cell Online 24(5):1860–1875

Burow MD, Chen Z-Y, Mouton TM, Moroney JV (1996) Isolation of cDNA clones of genes induced upon transfer of *Chlamydomonas reinhardtii* cells to low CO_2. Plant Mol Biol 31(2):443–448

Cardol P, González-Halphen D, Reyes-Prieto A, Baurain D, Matagne RF, Remacle C (2005) The mitochondrial oxidative phosphorylation proteome of *Chlamydomonas reinhardtii* deduced from the Genome Sequencing Project. Plant Physiol, 137 (2):447–459

Cardol P, Vanrobaeys F, Devreese B, Van Beeumen J, Matagne RF, Remacle C (2004) Higher plant-like subunit composition of mitochondrial complex I from *Chlamydomonas reinhardtii*: 31 conserved components among eukaryotes. Biochimica et Biophysica Acta (BBA)-Bioenergetics 1658 (3):212–224

Coragliotti AT, Beligni MVn, Franklin SE, Mayfield SP (2011) Molecular factors affecting the accumulation of recombinant proteins in the *Chlamydomonas reinhardtii* chloroplast. Mol Biotechnol 48 (1):60–75

Dent RM, Han M, Niyogi KK (2001) Functional genomics of plant photosynthesis in the fast lane using *Chlamydomonas reinhardtii*. Trends Plant Sci 6(8):364–371

Dodge JD (1968) The fine structure of chloroplasts and pyrenoids in some marine dinoflagellates. J Cell Sci 3(1):41–47

Duanmu D, Miller AR, Horken KM, Weeks DP, Spalding MH (2009b) Knockdown of a limiting-CO_2-inducible gene HLA3 decreases bicarbonate transport and photosynthetic C_i-affinity in *Chlamydomonas reinhardtii*. Proc Natl Acad Sci 106(14):5990–5995

Duanmu D, Wang Y, Spalding MH (2009a) Thylakoid lumen carbonic anhydrase (CAH3) mutation suppresses air-dier phenotype of LCIB mutant in *Chlamydomonas reinhardtii*. Plant Physiol 149(2):929–937

Eriksson M, Karlsson J, Ramazanov Z, Gardestram P, Samuelsson G (1996) Discovery of an algal mitochondrial carbonic anhydrase: molecular cloning and characterization of a low-CO_2-induced polypeptide in *Chlamydomonas reinhardtii*. Proc Natl Acad Sci 93(21):12031–12034

Eriksson M, Villand P, Gardestram P, Samuelsson G (1998) Induction and regulation of expression of a low-CO_2-induced mitochondrial carbonic anhydrase in *Chlamydomonas reinhardtii*. Plant Physiol 116(2):637–641

Evans JR, Von Caemmerer S (1996) Carbon dioxide diffusion inside leaves. Plant Physiol 110(2):339

Fan J, Yan C, Andre C, Shanklin J, Schwender J, Xu C (2012) Oil accumulation is controlled by carbon precursor supply for fatty acid synthesis in *Chlamydomonas reinhardtii*. Plant Cell Physiol 53(8):1380–1390

Fang W, Si Y, Douglass S, Casero D, Merchant SS, Pellegrini M, Ladunga I, Liu P, Spalding MH (2012) Transcriptome-wide changes in *Chlamydomonas reinhardtii* gene expression regulated by carbon dioxide and the CO_2-concentrating mechanism regulator CIA5/CCM1. The Plant Cell Online 24(5):1876–1893

Fujiwara S, Fukuzawa H, Tachiki A, Miyachi S (1990) Structure and differential expression of two genes encoding carbonic anhydrase in *Chlamydomonas reinhardtii*. Proc Natl Acad Sci 87(24):9779–9783

Fukuzawa H, Ishida S, Miyachi S (1991) cDNA cloning and gene expression of carbonic anhydrase in *Chlamydomonas reinhardtii*. Can J Bot 69(5):1088–1096

Galván A, Rexach J, Mariscal V, Fernández E (2002) Nitrite transport to the chloroplast in Chlamydomonas reinhardtii: molecular evidence for a regulated process. J Exp Bot 53 (370):845–853

Giordano M, Norici A, Forssen M, Eriksson M, Raven JA (2003) An anaplerotic role for mitochondrial carbonic anhydrase in *Chlamydomonas reinhardtii*. Plant Physiol 132(4):2126–2134

Grossman AR, Croft M, Gladyshev VN, Merchant SS, Posewitz MC, Prochnik S, Spalding MH (2007) Novel metabolism in *Chlamydomonas* through the lens of genomics. Curr Opin Plant Biol 10(2):190–198

Hanson DT, Franklin LA, Samuelsson G, Badger MR (2003) The *Chlamydomonas reinhardtii* cia3 mutant lacking a thylakoid lumen-localized carbonic anhydrase is limited by CO_2 supply to RuBisCO and not photosystem II function in vivo. Plant Physiol 132(4):2267–2275

Hoek C (1995) Algae: an introduction to phycology. Cambridge university press, Cambridge

Hoog JL, Lacomble S, Toatoole E, Hoenger A, McIntosh JR, Gull K (2014) Modes of flagellar assembly in *Chlamydomonas reinhardtii* and *Trypanosoma brucei*. eLife 3

Im CS, Grossman AR (2002) Identification and regulation of high light-induced genes in *Chlamydomonas reinhardtii*. Plant J 30(3):301–313

Im C-S, Zhang Z, Shrager J, Chang C-W, Grossman AR (2003) Analysis of light and CO_2 regulation in *Chlamydomonas reinhardtii* using genome-wide approaches. Photosynth Res 75(2):111–125

Jungnick N, Ma Y, Mukherjee B, Cronan JC, Speed DJ, Laborde SM, Longstreth DJ, Moroney JV (2014) The carbon concentrating mechanism in *Chlamydomonas reinhardtii*: finding the missing pieces. Photosynth Res: 1–15

Kaplan A, Reinhold L (1999) CO_2 concentrating mechanisms in photosynthetic microorganisms. Annu Rev Plant Biol 50(1):539–570

Karlsson J, Clarke AK, Chen Z, Hugghins SY, Park Y, Husic HD, Moroney JV, Samuelsson G (1998) A novel α-type carbonic anhydrase associated with the thylakoid membrane in *Chlamydomonas reinhardtii* is required for growth at ambient CO_2. The EMBO J 17(5):1208–1216

Kimpel DL, Togasaki RK, Miyachi S (1983) Carbonic anhydrase in *Chlamydomonas reinhardtii* I Localization. Plant Cell Physiol 24(2):255–259

Kustu S, Inwood W (2006) Biological gas channels for NH_3 and CO_2: evidence that Rh (Rhesus) proteins are CO_2 channels. Transfus Clin Biol 13(1):103–110

Ma Y, Pollock SV, Xiao Y, Cunnusamy K, Moroney JV (2011) Identification of a novel gene, CIA6, required for normal pyrenoid formation in *Chlamydomonas reinhardtii*. Plant Physiol 156(2):884–896

Maren TH (1967) Carbonic anhydrase: chemistry, physiology, and inhibition. Physiol Rev 47(4):595–781

Mariscal V, Moulin P, Orsel M, Miller AJ, Fernández E, Galván A (2006) Differential Regulation of the *Chlamydomonas Nar1* Gene Family by Carbon and Nitrogen. Protist 157 (4):421–433

Merchant SS, Prochnik SE, Vallon O, Harris EH, Karpowicz SJ, Witman GB, Terry A, Salamov A, Fritz-Laylin LK, Marachal-Drouard L (2007) The *Chlamydomonas* genome reveals the evolution of key animal and plant functions. Science 318(5848):245–250

Meyer MT, Genkov T, Skepper JN, Jouhet J, Mitchell MC, Spreitzer RJ, Griffiths H (2012) RuBisco small-subunit α–helices control pyrenoid formation in *Chlamydomonas*. Proc Natl Acad Sci 109(47):19474–19479

Mitra M, Lato SM, Ynalvez RA, Xiao Y, Moroney JV (2004) Identification of a new chloroplast carbonic anhydrase in *Chlamydomonas reinhardtii*. Plant Physiol 135(1):173–182

Mitra M, Mason CB, Xiao Y, Ynalvez RA, Lato SM, Moroney JV (2005) The carbonic anhydrase gene families of *Chlamydomonas reinhardtii*. Can J Bot 83(7):780–795

Miura K, Yamano T, Yoshioka S, Kohinata T, Inoue Y, Taniguchi F, Asamizu E, Nakamura Y, Tabata S, Yamato KT (2004) Expression profiling-based identification of CO_2-responsive genes regulated by CCM1 controlling a carbon-concentrating mechanism in Chlamydomonas reinhardtii. Plant Phys 135(3):1595–1607

Moroney JV, Husic HD, Tolbert NE (1985) Effect of carbonic anhydrase inhibitors on inorganic carbon accumulation by *Chlamydomonas reinhardtii*. Plant Physiol 79(1):177–183

Moroney JV, Ma Y, Frey WD, Fusilier KA, Pham TT, Simms TA, DiMario RJ, Yang J, Mukherjee B (2011) The carbonic anhydrase isoforms of *Chlamydomonas reinhardtii:* intracellular location, expression, and physiological roles. Photosynth Res 109(1–3):133–149

Moroney JV, Ynalvez RA (2007) Proposed carbon dioxide concentrating mechanism in *Chlamydomonas reinhardtii*. Eukaryot Cell 6(8):1251–1259

Mukherjee B (2013) Investigation of the role of putative inorganic carbon transporters in the carbon dioxide concentrating mechanisms of *Chlamydomonas reinhardtii*. Calcutta University, Calcutta

Nassoury N, Fritz L, Morse D (2001) Circadian changes in ribulose-1, 5-bisphosphate carboxylase/oxygenase distribution inside individual chloroplasts can account for the rhythm in dinoflagellate carbon fixation. The Plant Cell Online 13(4):923–934

Ohnishi N, Mukherjee B, Tsujikawa T, Yanase M, Nakano H, Moroney JV, Fukuzawa H (2010) Expression of a Low CO_2–inducible protein, LCI1, increases inorganic carbon uptake in the green alga *Chlamydomonas reinhardtii*. The Plant Cell Online 22(9):3105–3117

Palenik B, Grimwood J, Aerts A, Rouzac P, Salamov A, Putnam N, Dupont C, Jorgensen R, Derelle E, Rombauts S (2007) The tiny eukaryote *Ostreococcus* provides genomic insights into the paradox of plankton speciation. Proc Natl Acad Sci 104(18):7705–7710

Park Y-I, Karlsson J, Rojdestvenski I, Pronina N, Klimov V, Ã–quist G, Samuelsson Gr (1999) Role of a novel photosystem II-associated carbonic anhydrase in photosynthetic carbon assimilation in *Chlamydomonas reinhardtii*. FEBS Lett 444 (1):102–105

Peers G, Niyogi KK (2008) Pond scum genomics: The genomes of *Chlamydomonas* and *Ostreococcus*. The Plant Cell Online 20(3):502–507

Pollock SV, Prout DL, Godfrey AC, Lemaire SD, Moroney JV (2004) The *Chlamydomonas reinhardtii* proteins Ccp1 and Ccp2 are required for long-term growth, but are not necessary for efficient photosynthesis, in a low-CO_2 environment. Plant Mol Biol 56(1):125–132

Price GD, Howitt SM, Harrison K, Badger MR (1993) Analysis of a genomic DNA region from the cyanobacterium *Synechococcus* sp. strain PCC7942 involved in carboxysome assembly and function. J Bacteriol 175(10):2871–2879

Ramazanov Z, Rawat M, Henk MC, Mason CB, Matthews SW, Moroney JV (1994) The induction of the CO_2-concentrating mechanism is correlated with the formation of the starch sheath around the pyrenoid of *Chlamydomonas reinhardtii*. Planta 195(2):210–216

Raven JA (1997) CO_2-concentrating mechanisms: a direct role for thylakoid lumen acidification? Plant Cell Environment 20(2):147–154

Raven JA, Johnston AM, Kobler JE, Korb R, McInroy SG, Handley LL, Scrimgeour CM, Walker DI, Beardall J, Clayton MN (2002) Seaweeds in cold seas: evolution and carbon acquisition. Ann Bot 90(4):525–536

Rawat M, Henk MC, Lavigne LL, Moroney JV (1996) *Chlamydomonas reinhardtii* mutants without ribulose-1, 5-bisphosphate carboxylase-oxygenase lack a detectable pyrenoid. Planta 198(2):263–270

Rawat M, Moroney JV (1995) The regulation of carbonic anhydrase and ribulose-1, 5-bisphosphate carboxylase/oxygenase activase by light and CO_2 in *Chlamydomonas reinhardtii*. Plant Physiol 109(3):937–944

Rolland N, Dorne A, Amoroso G, Saltemeyer DF, Joyard J, Rochaix JD (1997) Disruption of the plastid *ycf10* open reading frame affects uptake of inorganic carbon in the chloroplast of *Chlamydomonas*. The EMBO J 16(22):6713–6726

Schornstein KL, Scott J (1982) Ultrastructure of cell division in the unicellular red alga *Porphyridium purpureum*. Can J Bot 60(1):85–97

Shevela D, Eaton-Rye JJ, Shen J-R (2012) Photosystem II and the unique role of bicarbonate: A historical perspective. Biochimica et Biophysica Acta (BBA)-Bioenergetics 1817 (8):1134-1151

Shutova T, Kenneweg H, Buchta J, Nikitina J, Terentyev V, Chernyshov S, Andersson B, Allakhverdiev SI, Klimov VV, Dau H (2008) The photosystem II associated Cah3 in *Chlamydomonas* enhances the O2 evolution rate by proton removal. The EMBO J 27(5):782–791

Sinetova MA, Kupriyanova EV, Markelova AG, Allakhverdiev SI, Pronina NA (2012) Identification and functional role of the carbonic anhydrase Cah3 in thylakoid membranes of pyrenoid of *Chlamydomonas reinhardtii*. Biochimica et Biophysica Acta (BBA)-Bioenergetics 1817 (8):1248–1255

Sonoda M, Katoh H, Vermaas W, Schmetterer G, Ogawa T (1998) Photosynthetic electron transport involved in *Pxc*A-dependent proton extrusion in *Synechocystis* sp. strain PCC6803: effect of *pxc*A inactivation on CO_2, HCO_3^-, and NO_3^- uptake. J Bacteriol 180 (15):3799–3803

Soupene E, Inwood W, Kustu S (2004) Lack of the Rhesus protein Rh1 impairs growth of the green alga *Chlamydomonas reinhardtii* at high CO_2. Proc Nat Acad Sci United States of America 101(20):7787–7792

Spalding MH (2008) Microalgal carbon-dioxide-concentrating mechanisms: *Chlamydomonas* inorganic carbon transporters. J Exp Bot 59(7):1463–1473

Spalding MH, Jeffrey M (1989) Membrane-associated polypeptides induced in *Chlamydomonas* by limiting-CO_2 concentrations. Plant Physiol 89(1):133–137

Trippens J, Greiner A, Schellwat J, Neukam M, Rottmann T, Lu Y, Kateriya S, Hegemann P, Kreimer G (2012) Phototropin influence on eyespot development and regulation of phototactic behavior in *Chlamydomonas reinhardtii*. The Plant Cell Online 24(11):4687–4702

Tsuzuki M, Miyachi S (1989) The function of carbonic anhydrase in aquatic photosynthesis. Aquat Bot 34(1):85–104

Van K, Spalding MH (1999) Periplasmic carbonic anhydrase structural gene (Cah1) mutant in *Chlamydomonas reinhardtii*. Plant Physiol 120(3):757–764

Villand P, Eriksson M, Samuelsson G (1997) Carbon dioxide and light regulation of promoters controlling the expression of mitochondrial carbonic anhydrase in *Chlamydomonas reinhardtii*. Biochem J 327:51–57

Wang Y, Duanmu D, Spalding MH (2011) Carbon dioxide concentrating mechanism in *Chlamydomonas reinhardtii*: inorganic carbon transport and CO_2 recapture. Photosynth Res 109(1–3):115–122

Wang Y, Spalding MH (2006) An inorganic carbon transport system responsible for acclimation specific to air levels of CO_2 in *Chlamydomonas reinhardtii*. Proc Natl Acad Sci 103(26):10110–10115

Wang Y, Sun Z, Horken KM, Im C-S, Xiang Y, Grossman AR, Weeks DP (2005) Analyses of CIA5, the master regulator of the carbon-concentrating mechanism in *Chlamydomonas reinhardtii*, and its control of gene expression. Can J Bot 83(7):765–779

Xiang Y, Zhang J, Weeks DP (2001) The Cia5 gene controls formation of the carbon concentrating mechanism in *Chlamydomonas reinhardtii*. Proc Natl Acad Sci 98(9):5341–5346

Yamano T, Fukuzawa H (2009) Carbon concentrating mechanism in a green alga, *Chlamydomonas reinhardtii*, revealed by transcriptome analyses. J Basic Microbiol 49(1):42–51

Yamano T, Tsujikawa T, Hatano K, Ozawa S-i, Takahashi Y, Fukuzawa H (2010) Light and low-CO_2-dependent LCIB-LCIC complex localization in the chloroplast supports the carbon-concentrating mechanism in *Chlamydomonas reinhardtii*. Plant Cell Physiol 51(9):1453–1468

Ynalvez RA, Xiao Y, Ward AS, Cunnusamy K, Moroney JV (2008) Identification and characterization of two closely related β-carbonic anhydrases from *Chlamydomonas reinhardtii*. Physiol Plant 133(1):15–26

Yoshihara C, Inoue K, Schichnes D, Ruzin S, Inwood W, Kustu S (2008) An Rh1-GFP fusion protein is in the cytoplasmic membrane of a white mutant strain of *Chlamydomonas reinhardtii*. Mol Plant 1(6):1007–1020

Yoshioka S, Taniguchi F, Miura K, Inoue T, Yamano T, Fukuzawa H (2004) The novel Myb transcription factor LCR1 regulates the CO_2-responsive gene Cah1, encoding a periplasmic carbonic anhydrase in *Chlamydomonas reinhardtii*. The Plant Cell Online 16(6):1466–1477

Chapter 5
Photosynthetic Microorganism-Based CO$_2$ Mitigation System: Integrated Approaches for Global Sustainability

5.1 Introduction

One of the most urgent challenges facing humanity today is to conserve and sustain natural resources, including air, water, and fossil fuels, at the same time protecting the environment. Among all pressing challenges, global warming is one of the biggest threats facing the planet (Gore 2006). It is the raise in planet's standard surface temperature due to sun heat-trapping ability of greenhouse gases (GHG$_s$), emitted from various anthropogenic sources. Carbon dioxide (CO$_2$) is the largest contributing gas to green house effect. The CO$_2$ mitigation refers to various approaches to prevent or reduce emission of CO$_2$ gas by anthropogenic sources and improve their removal from environment. Efforts underway around the world, range from ideas to remove CO$_2$ from the atmosphere (air CO$_2$ capture) and flue gases (carbon capture and storage), and prevent carbon in biomass from reentering the atmosphere, such as with bioenergy with carbon capture and storage (BECCS). In this context, photosynthetic microorganism-based carbon mitigation system could provide an efficient photosynthetic pathways to reduce the excess atmospheric CO$_2$ concentrations and balanced the O$_2$ and CO$_2$ concentrations in atmosphere. Owing to wide range of applications of photosynthetic microorganism-based carbon mitigation system, this chapter addresses the all possible synergistic benefits from integrated sustainable approaches of algal technology and offers key insights that can be used to mitigate carbon dioxide pollution.

5.2 Necessity of Sustainable System for Carbon Mitigation

Across the world, industrialization paved the way for profound technological advancement which helps to provide better health care, employment, education, infrastructure, and social life (Mollenkopf and Fozard 2003). At the same time, rapid industrialization also facilitated the resource depletion and ruin of local ecological system with pollution (Wackernagel and Rees 1998). For their

© The Author(s) 2014
S.K. Singh et al., *Photosynthetic Microorganisms*, SpringerBriefs in Materials,
DOI 10.1007/978-3-319-09123-5_5

energy need, industries are relied on the finite resources of non-renewable fossil fuels (coal, natural gas, and oil). However, fossil fuel combustion leads to increase emission of GHGs such as carbon dioxide (CO_2), sulfur dioxide (SO_2), and nitrogen oxides (NO_x) (Miller and Spoolman 2008). Although, primary greenhouse gas CO_2 is naturally occur in the Earth's atmosphere as a part of the carbon cycle (Pearce 1992). However, by adding more CO_2 and influencing the natural sinks such as forests, human activities are constantly altering the dynamic balance of normal carbon cycle. The paleoclimate records showed the strong correspondence between temperature and the atmospheric carbon dioxide concentration, during the glacial cycles of the past several hundred thousand years. Averaging 80 proxy temperature records over the most recent deglaciation indicates that global mean temperature is highly correlated and varied nearly in phase with rise in CO_2, suggested a primary role for CO_2 in driving global warming and deglaciation (Shakun et al. 2012). In fact, CO_2 is a heat-trapping gas which enhances the atmosphere's ability to trap sun heat, pushing the world into dangerous territory. As a major sink for atmospheric CO_2, oceans are also no longer able to absorb the increased concentration of CO_2 in the atmosphere without changes to the acidity levels. Since the dawn of the industrial revolution, carbon emissions from burning fossil fuels have grown exponentially. Despite wide carbon limit agreement by governments, CO_2 emissions have grown almost 2.6 % each year, hitting an all-time high of 9.7 billion tons of carbon in 2012, a grim new milestone on the path of climate disruption (Olivier et al. 2012). According to the latest on-site measurements by the Scripps Institute of Oceanography and statistical data of Earth Policy Institute, global atmospheric CO_2 concentrations raised progressively faster year by year; 396.48 parts per million (ppm) in 2013, up from 389.86 ppm in 2010 and from 290 ppm in 1880 (Dlugokencky et al. 2013; Wang et al. 2014) (shown in Fig. 5.1). Correspondingly, according to NASA data (NASA 2014), global temperature in 2013 averaged 58.3 °F, roughly a degree warmer than the twentieth-century

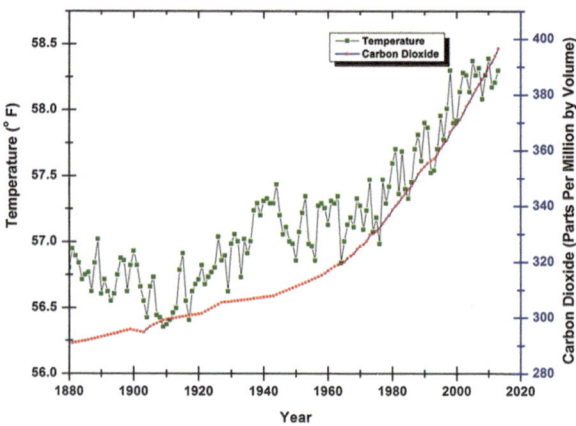

Fig. 5.1 Global temperature and carbon dioxide concentration (1880–2013)

average. Concentration of methane (CH_4) and nitrous oxide (N_2O) have also risen; however, they account for a smaller share of GHGs, 17 and 8.7 %, respectively (Hockstad and Cook 2012). Emissions of CO_2 are likely to increase significantly in the near future due to increased energy demand from developing countries. The continual rise in anthropogenic atmospheric CO_2 and the concomitant effect on climate change underlie the urgent need for the implementation of carbon mitigation approaches to alleviate the serious global climate repercussions. The Kyoto Protocol of 1997 called for a 5.2 % reduction in GHG emissions worldwide from 1990 values (Jaggi et al. 2011). To meet the agreed target, a number of efforts have been directed to explore a range of effective technologies to reduce CO_2 emissions, including chemical reaction-based approaches, direct injection to underground or to the ocean and biological CO_2 mitigation possibilities. Chemical reaction-based approaches consist of washing with alkaline solutions (Diao et al. 2004), multiwalled carbon nanotubes (Su et al. 2009), amine-coating activated carbon (Plaza et al. 2007), etc. Direct injection method popularly known as carbon capture and storage (CCS) methodologies covers three steps: CO_2 capture from flue gases, its transportation and storage through deep-ocean injection, and mineralization (Davies et al. 2013). However, both technologies not only are expensive (use so much inputs and energy), but also have unsustainable (store enormous amount of flue gases) option, makes mitigation benefits marginal. These procedures should only be considered as short-term solutions. The overall implication is therefore a need for enhancement of global strategies for mitigation of CO_2-related emissions and development of CO_2-neutral energy resources. In this scenario, biological CO_2 mitigation in which CO_2 being biologically transformed to valuable organic products seems to be cost-effective and sustainable approach. Biological CO_2 fixation is currently achieved through the photosynthesis of all terrestrial plants and a tremendous number of photosynthetic microorganisms. Globally, 500 billion tonnes CO_2 is fixed annually by terrestrial vegetation which is 20 times more than the amount of CO_2 released from fossil fuel consumption (Skjanes et al. 2007). However, photosynthetic microorganisms can grow much faster than terrestrial plants, and their CO_2-fixation efficiency is about 10–50 times better (Goswami and Kalita 2012). Photosynthetic microorganisms are prokaryotes (cyanobacteria) or simple unicellular eukaryotes (microalgae) that can grow rapidly and live in harsh conditions due to their structure.

5.3 Perspectives of Photosynthetic Microorganism-Based CO_2 Mitigation

Photosynthesis is the nature's way to recycle CO_2 (Acton 2013). In the perspective of challenges associated with existing CO_2 mitigation technologies, photosynthetic microorganism-based CO_2 mitigation is gaining wide attention as an alternative renewable source of biomass for the production of CO_2-neutral energy resources. Photosynthetic microorganisms fix the C_i from flue gas by two routes: conversion into biomass via photosynthesis and calcium carbonate production through calcification process (Buitenhuis et al. 1999). They are the most productive transformers of CO_2 into O_2 and actively play a significant role in the creation of organic biomass and oxygenic environment on Earth (Mostafa and El-Gendy 2013). If we assume carbon mass fraction of the algae is half of their dry biomass, then to produce 1 kg dry algal biomass, we need at least 1.83 kg CO_2 (Slade and Bauen 2013). However, in reality, carbon content is almost more to this, as shown in Table 5.1.

Thus, CO_2 from industrial flue gases would be willingly biofix into organic biomass. This organic biomass can be later transformed into valuable products such as bioethanol, biofuel, hydrogen, amino acids, proteins, fatty acids, vitamin A, minerals, pigments, dietary supplements for human, animals and aquaculture, and other biocompounds. (Pulz and Gross 2004; Skjanes et al. 2007; Cuellar-Bermudez et al. 2014). Some of the key advantages include the following: High-purity CO_2 gas is not required for cultivation (Olaizola 2003a, b), other polluting products such as NO_x can be used as nutrients (Singh et al. 2014a), and some species able to grow in high and in below to freezing temperatures (Castenholz 1969; Singh et al. 2014b, c) yield high-value commercial products with a renewable cycle. Over the past few years, several studies have highlighted the potential of photosynthetic microorganism species to reduce CO_2 emissions (Toledo-Cervantes et al. 2013; Yoo et al. 2010; Chiang et al. 2011). Several microalgal species have shown good tolerance to sparging with gas containing 5–20 % CO_2, i.e., concentrations as in flue gas. Toledo-Cervantes et al. (2013) found CO_2 fixation rate of 970 g m^{-3} d^{-1} in *Scenedesmus obtusiusculus*, grown under 10 % of CO_2. Correspondingly, Ramaraj et al. (2010) evaluated the CO_2 fixation rate in four microalgae and reported *Botryococcus braunii* has the highest fixation capability of 496.98 mg L^{-1} day^{-1}, followed by *Spirulina platensis, Dunaliella tertiolecta,* and *Chlorella vulgaris* (318.61, 272.4, and 251.64 mg L^{-1} day^{-1}, respectively). In a recent study, Ramaraj et al. (2014) showed that freshwater microalgal biomass is solely derived from atmospheric CO_2. H-CO_2 concentration may also stimulate growth of microalgae to faster

Table 5.1 Carbon content of some photosynthetic microorganisms

Species	Carbon content (dry weight %)	References
Dunaliella teritolecta SAG-13.86	36	Sydney et al. (2010)
Chlorogleopsis sp.	41	Ono and Cuello (2007)
Chlorella vulgaris LEB-104	45	Sydney et al. (2010)
Spirulina plantesis LEB-52	50	Sydney et al. (2010)
Chlorella pyrenoidosa SJTU-2	49–51	Tang et al. (2011)
Scenedesmus obliquus SJTU-3	49–51	Tang et al. (2011)
Chlorella sp. UK001	54	Hirata et al. (1996)
Botrycoccus braunii SAG-30.81	58	Sydney et al. (2010)
Euglena gracilis	65	Chae et al. (2006)

compare to atmospheric air. Jiang et al. (2011) found that specific growth rate of *Nannochloropsis* sp. increased around 58 % from 0.33 to 0.52 d^{-1} with shift in atmospheric air aeration to 15 % CO_2. This is a plausible observation because high concentration of CO_2 promotes photosynthetic efficiency of microalgae to reproduce within a shorter time, and thus, more quantity of microalgae biomass could be attained. However, under high CO_2 conditions, CCMs may also play an important role by activate carbonic anhydrases to prevent inhibition of RuBisCO enzyme due to carbonic acidification (Solovchenko and Khozin-Goldberg 2013). Thus, acidification is the major factor to inhibit microalgal growth in H-CO_2 environment. In certain microalgal species, the addition of bases starts to compensate for CO_2 acidification. This enhances the CO_2 tolerance of some microalgal species to sustain even at 100 % CO_2 (Olaizola 2003a, b). Overall, the CO_2 tolerance of any microalgal strain is dependent on important parameters such as strains (Tang et al. 2011), nutritional condition, light (Soletto et al. 2008), cell density (Chiu et al. 2008), and pH of medium (Olaizola 2003a, b).

5.4 Sustainable Integration of Photosynthetic Microorganism-Based CO_2 Mitigation System with Industries

Sustainability is the ways of exploiting the natural resource, under which humans and nature can coexist with fulfilling the environmental and socioeconomic requirements of present and future generations (Flint 2013). In order to achieve full processing capabilities and develop an effective photosynthetic microorganism-based CO_2 mitigation system, recent research efforts have concentrated on evaluating and optimizing the integrated multidisciplinary system. Through industrial symbiosis of CO_2 mitigation system with photosynthetic microorganism cultivation and exploiting the produced biomass in various sectors (biofuel, environment, agriculture, food, pharma, etc.), which are usually wasted, the coupled system provides an economically feasible and environmentally sustainable model for the successful CO_2 mitigation (Skjanes et al. 2007) (see Fig. 5.2).

5.4.1 Biofuel Industry

The biofuel industry is one of the growing sub-sectors of the global energy sector. The report from pike research predicts the global biofuels market will more than double over the next decade, increasing from $82.7 billion in 2011 to $185.3 billion in 2021 (Martinot 2013; Lin 2014). The biofuel is referred to as efficient substitute (solid, liquid, or gaseous fuels) of fossil fuels, predominantly derived from biorenewable feedstocks which contain energy from geologically recent carbon fixation. They are associated with widespread availability, affordability, accessibility of

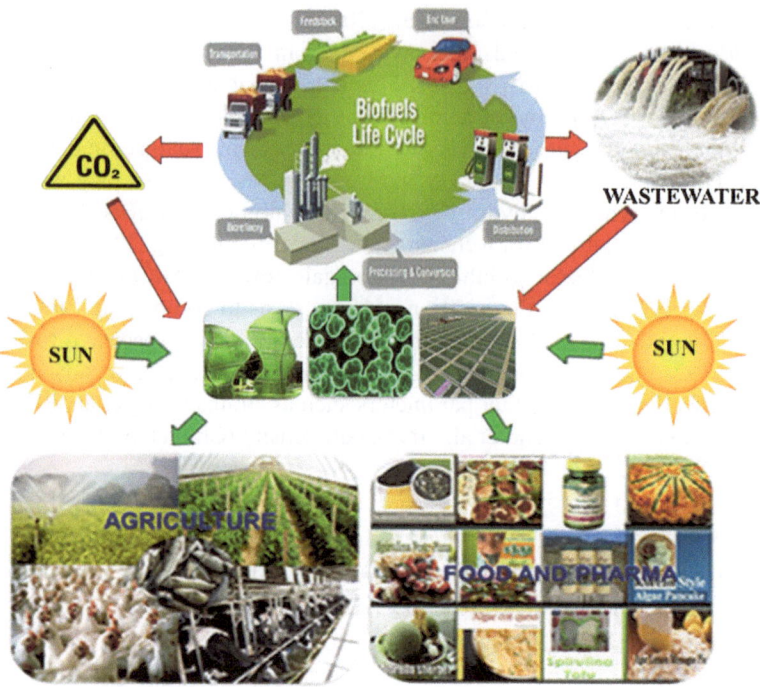

Fig. 5.2 Industrial integration of photosynthetic microorganism-based CO$_2$ mitigation system

technology, ease of transport, storage, versatility in use in engines, and socioeconomic and environmental benefits (Costa and De Morais 2011). However, the rapid
increase in production of biofuel feedstock may threaten biodiversity. The sustainability of biofuel depends upon the growing practices of eco-friendly feedstocks,
smallest ecological footprints during cultivation and ability to sequester carbon or
have a negative or zero carbon balance over the entire production life cycle (Groom
et al. 2008). Unfortunately, sources of most common biofuel (bioethanol and biodiesel) are traditional edible oil crops such as palm, cone, soy, used as food also.
Hence, converting them as fuel has an impact on food availability and food prices
which may even cause future food scarcity.

The best alternatives appear to be fuels of the future, especially fuels derived
from photosynthetic microorganisms. They are believed to be the first group of
organisms on Earth to perform photosynthesis. In recent years, they have received
notable in both academic and commercial biofuels research. The advantages of
using photosynthetic microorganism to produce biofuels are their simple cell division cycle, abundance almost everywhere on Earth, acquisition of valuable compounds through photosynthesis, tolerance to varying environmental conditions
such as grow in bare land with waste or brackish water, nonstop production, and
ease to be genetically engineered. (González-López et al. 2012). One of the great
things about photosynthetic microorganism-based fuels is that the resources available to produce them; CO$_2$, water, and sunlight are practically unlimited. Thus,

they can also help in reducing the main greenhouse gas, CO_2. Groom et al. (2008) compared the GHGs emissions of various existing biofuel feedstocks during cultivation, harvesting, and fuel production (see Table 5.2). They reported that algae biodiesel GHGs emission was extremely less compared to other biofuel feedstock crops. Similarly, the total area required for oil of biodiesel production is also significantly smaller from species in use today. The amount of land needed for the corresponding production using microalgae would be around 100–200 times less. Current estimates on existing technologies show that algae farms can produce up to 15,000 gallons of oil per acre per year which substantially reduces land usage for biodiesel feedstock. Thus, algal biomass has the potential to be converted into biofuel, yielding a CO_2-neutral energy carrier, i.e., the carbon it takes in will be released again when it is burned (Singh et al. 2014a; Jones and Mayfield 2012).

As shown Fig. 5.3, algal biomass can be converted into fuel by thermochemical (Bridgwater 2003; Tanger et al. 2013) or biochemical (Sreekrishnan et al. 2004) routes. Thermochemical methods include gasification (converting the carbonaceous biomass into synthesis gas (syngas) such as methane and H_2), liquefaction (converting biomass into liquid hydrocarbon), pyrolysis (producing fuel gases, oils, and charcoal), or hydrogenation (addition of H_2 to saturate organic compounds). Biochemical conversion methods include fermentation of the biomass by bacteria to produce energy carriers such as bioethanol (John et al. 2011), biohydrogen (Singh and Olsen 2011), and biomethane (Iyovo et al. 2010), or extraction of oils from the biomass for biodiesel production (Sander and Murthy, 2010).

All form of biofuel production will re-emit CO_2, which can be recaptured from algal biofuel plants. It is a technology that marries the potential need for carbon disposal in the electric utility industry with the need for clean-burning alternatives to petroleum in the transportation sector. The biofuel productivity of some strains has shown in the Table 5.3.

5.4.1.1 Biodiesel Fuel

Algal biodiesel is an exciting possibility existing for recycles of huge amount of CO_2 produced in existing fossil fuel power plants. It will ultimately leads to reduction of harmful emissions of carbon monoxide, hydrocarbons, and particulate matter and to the elimination of SO_x emissions. It is reported that 0.02 ± 0.004 t of CO_2 are required for making per gallon of biodiesel from algae (Sheehan et al. 1998). Biodiesel is defined as non-petroleum-based diesel fuel consisting of alkyl esters (mainly methyl, but also ethyl, and propyl) of long-chain fatty acids, produced by conversion of lipid in the presence of an acid or basic catalyst through transesterification reaction with glycerol as a by-product (Leung et al. 2010). To meet the required demand and add 5.0 % biodiesel (B5) to mineral diesel oil, one would have to increase the production of vegetable oils by 50–100 %. This is a difficult goal to achieve with these oils alone, since it represents a proportional increase in arable land with oil crops, and the current agricultural productivity has reached values that are difficult to increase. However, microalgae can double

Table 5.2 A comparative data of various biofuel feedstock

Feedstock crop	Biomass (Mt/ha/year) Groom et al. (2008)	GHGs emission[a] Groom et al. (2008)	Oil content (% dry mass) Groom et al. (2008)	Oil yield (L ha^{-1}) (Chisti et al. 2008)	Gallons of oil per acre per year http://oakhavenpc.org/ cultivating_algae.htm	Biodiesel (Mt/ha/year)	Ethanol yield (gal/acre) (Mussatto et al. 2010)	Ethanol yield (L/ha) (Mussatto et al. 2010)	Energy content (boe/1,000 ha/day)
Corn	–	81–85	–	172	18	2.2–5.3	370–430	3,460–4,020	–
Corn stover	–	–	–	–	–	–	112–150	1,050–1,400	–
Coconut	–	–	–	2,689	287	–	–	–	–
Jatropha	7.5–10	–	30–50	1,892	207	–	–	–	40–100
Wheat	–	–	–	–	–	–	277	2,590	–
Cassava	–	–	–	–	–	–	354	3,310	–
Sweet sorghum	–	–	–	–	–	–	326–435	3,050–4,070	–
Switch grass	–	–24	–	–	–	–	1,150	10,760	–
Sugar cane	–	4–12	–	–	–	–	536–714	5,010–6,680	–
Soybeans	1–2.5	49	20	446	48	0.2–0.5	–	–	3–8
Rapeseed oil	3	37	40	–	127	1.2	–	–	22
Palm oil	19	N/A	20	5,950	635	3.7	–	–	63
Algae	140–255	–183	30–70	58,700–136,900	5,000–15,000	50–100	5,000–15,000	46,760–140,290	1,150–2000

[a]Kg of CO_2 created per mega joule of energy produced

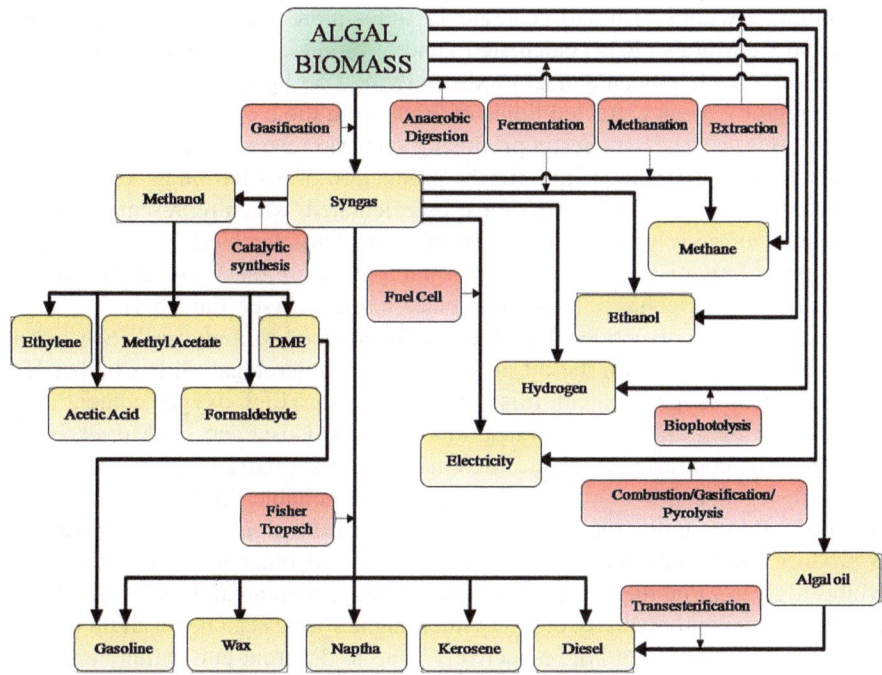

Fig. 5.3 Paths to several energy products from algal biomass

Table 5.3 Biofuel productivity of photosynthetic microorganisms

Strain	Biofuel	Cultivation method	Biofuel productivity	References
Spirulina platensis	Hydrogen	Erlenmeyer flask	1.8 μmol mg^{-1}	Aoyama et al. (1997)
Dunaliella sp.	Ethanol	Erlenmeyer flask	11.0 mg g^{-1}	Shirai et al. (1998)
Chlamydomonas reinhardtii	Hydrogen	Roux bottle	2.5 mL h^{-1}	Ghirardi et al. (2000)
Spirulina Leb 18	Methane	Open raceway	0.79 g L^{-1}	Costa et al. (2008)
Neochlorosis oleabundans	Biodiesel	Polyethylene bags	56.0 g g^{-1}	Gouveia and Oliveira (2009)
S. platensis UTEX 1926	Methane	Glass sealed bottle	0.40 m^3 kg^{-1}	Converti et al. (2009)
Chlorococum sp.	Ethanol	Photobioreactor bag	3.83 g L^{-1}	Harun et al. (2010a)
Chlorococum sp.	Biodiesel	Photobioreactor bag	10.0 g L^{-1}	Harun et al. (2010a)

their biomass within 24 h and generate 15 times more oil/acre than other plants. Costa and De Morais (2011) calculated yield of biodiesel from microalgae will be 98.4 m^3 ha^{-1} with a mean annual productivity of 1.53 kg m^{-3} day^{-1} (for tropical climate region), mean extracted lipids of 30 % from the biomass, and total area of around 123.0 m^3. They suggest a promising scenario (even if the concentration of lipids is 15 % of dry weight) that for the production of 5.4 billion m^3 of biodiesel, an area of approximately 5.4 Mha must be calculated, which represents only 3 % of the area currently used for the cultivation of plants for biodiesel. As the concentration of fatty oil and productivity of microalgae are much higher than that of plants, algal biodiesel has potential to replace fossil fuel (see Table 5.4).

Algal oil is a triglyceride, three fatty acids attached to a glycerol (Silva et al. 2012). In comparison with the oil of higher plants, microalgal oil has higher H/C ratio and lower oxygen content. It is important because high oxygen content is not attractive for the production of transportation fuels (Costa and De Morais 2011). Consequently, contrasted to plant, microalgal biodiesel has a high calorific value, low viscosity, and low density, make them more suitable for biofuel (Miao and Wu 2004). Furthermore, the feedstock oil derived from algae can be hydro-treated to produce bioalkanes such as gasoline or JP-8 and other jet fuels. Microalgae exhibit a great variability in lipid content. Among microalgae species, lipid content and biomass productivity of many microalgae species can be reach up to 80 % (*Botryococcus*) and 7.3 g/l/d, respectively. Due to different cultivation conditions and extraction methods of fatty acids, variations in oil content are also possible but levels of 20–50 % are quite common (Powell and Hill 2009; Suali and Sarbatly 2012).

A highly integrated microalgal biodiesel system, synergistically coupled with carbon sequestration from flue gas and coproduction of valuable products, can be cost effective under favorable market conditions (Kumar et al. 2010a, b). However, the cost of algae-derived biodiesel is proportional to the species-specific efficiency of algae to sequester carbon dioxide as lipids (Jones and Mayfield 2012). It is interesting to note that although microalgae are able to biofix CO_2 efficiently, however, reports on the effect of CO_2 on lipid content and composition are contradictory. Few studies reported that lipid content of photosynthetic microorganisms increases linearly with CO_2 concentration. Muradyan et al. (2004) shown that in unicellular halophilic green alga *Dunaliella salina* (known to be susceptible to CO_2 stress), total fatty acid content enhanced up to 30 % on the dry weight basis, as CO_2 concentration was increased 2–10 %. Some studies reported the increment of lipid content in microalgae cells was only 1–6 % although 10–15 % of CO_2 was supplied as carbon source (Yoo et al. 2010) (see Table 5.5).

Based on several studies, it is the possibility that C_i may be utilized in the other biological form such as protein, sugars, and pigments. Relatively high composition of these products was attained compared to lipid composition in *C. vulgaris*, *S. platensis*, *B. braunii*, and *D. tertiolecta* (Sydney et al. 2010). Many studies reported that photosynthetic microorganism's lipid content is high in unsaturated fatty acids, an added advantage for biodiesel. Unsaturated fatty acids reduce the pour point of biodiesel and making them feasible to be used in temperate

Table 5.4 Oil content of photosynthetic microorganisms

Photosynthetic microorganism	Oil content (% dry wt) (Chisti 2007; Li et al. 2008; Mata et al. 2010; Sialve et al. 2009; Um and Kim 2009).
Ankistrodesmus sp.	24–31
Botryococcus braunii	25–75
Chlorella sp.	10–48; 28–32
Chlorella emersonii	25–63
Chlorella minutissima	57
Chlorella protothecoides	14–57
Chlorella sorokiniana	19–22
Chlorella vulgaris	5–58
Chaetoceros muelleri	33
Chlamydomonas reinhardtii	21
Crypthecodinium cohnii	20–51
Cylindrotheca sp.	16–37
Dunaliella sp.	17–67
Dunaliella primolecta	23
Dunaliella salina	6–25
Dunaliella tertiolecta	16–71
Euglena gracilis	14–20
Ellipsoidion sp.	27
Haematococcus pluvialis	25
Isochrysis galbana	7–40
Isochrysis sp.	7–33
Monodus subterraneus	16
Monallanthus salina	20–22
Nannochloris sp.	20–35; 20–56; 12–53
Nannochloropsis oculata	22–29; 31–68
Neochloris oleoabundans	29–65, 35–54
Nitzschia sp.	45–47
Pavlova salina	30
Pavlova lutheri	35
Phaeodactylum tricornutum	20–30; 18–57
Prostanthera incisa	62
Prymnesium parvum	22–39
Pyrrosia laevis	69.1
Schizochytrium sp.	50–77
Scenedesmus obliquus	11–55
Scenedesmus dimorphus	16-40
Schizochytrium sp.	50–77
Skeletonema costatum	13–51
Tetraselmis suecica	15–23
Thalassiosira pseudonana	20
Zitzschia sp.	45–47

Table 5.5 Fatty content of photosynthetic microorganisms grown under various CO_2 concentrations

Photosynthetic microorganism	CO_2 concentrations (%)	Nitrogen starvation	Fatty acid (%)	References
Dunaliella salina	10	No	30	Muradyan et al. (2004)
Chlorococcum littorale	5	Yes	34	Ota et al. (2009)
S. obliquus CNW-N	2.5	Yes	22.4	Ho et al. (2012)
Ettlia sp. YC001	5	No	42	Yoo et al. (2013)
Scenedesmus obliquus	0.04 5	Yes	15 55.7	Toledo-Cervantes et al. (2013)

countries. Tsuzuki et al. (1990) confirmed that degree of unsaturation of fatty acids significantly increased in *C. vulgaris* after shifting the 11-h-grown cells with air (L-CO2 cells) to the air enriched with 2 % CO_2 (H-CO2 cells). On the other hand, it is interesting to find that increasing the CO_2 concentration to cultivate some microalgae tends to favor the accumulation of total lipids and polyunsaturated fatty acid in the microalgae cells. For example, by supplying 50 % CO_2 to cultivate *Scenedesmus obliquus*, the lipid extracted from the microalgae consisted of 76 % unsaturated fatty acid, mainly linoleic acid (C18:2) and linolenic acid (C18:3) (Tang et al. 2011). This increase in polyunsaturated fatty acids has been explained by a relative low concentration of oxygen in water that might affect enzymatic desaturation metabolism of microalgae and therefore increasing the polyunsaturated fatty acid content in microalgae cells (Van Den Hende et al. 2012). It has also been observed that fatty acids' desaturation occurs more rapidly when CCM is induced in *Chlamydomonas reinhardtii* (Pronina et al. 1998). Another hypothesis confirmed by many researchers is that lipid accumulation rate enhanced with nitrogen starvation condition and increase in HCO_3^-/CO_2 ratio. Ota et al. (2009) found that at CO_2 concentrations in between 5 and 50 % under nitrate-rich environment, the total fatty acid content of *Chlorococcum littorale* was almost constant. After nitrate depletion, the content drastically increased with a decrease in CO_2 concentration. It seems nitrate limitation and HCO_3^-/CO_2 ratio in the culture media may be a controlling factor for fatty acid production. Maximum fatty acid content of 34 wt.% was observed at 5 % CO_2 concentration. Toledo-Cervantes et al. (2013) also reported that at 5 % CO_2 and without any nutrient limitation, *S. obtusiusculus* contained 20 % of nonpolar lipids, enhanced to 55.7 % under nitrogen starvation. Thus, two-stage cultivation process is suggested to enhance the lipid content in photosynthetic microorganism cells which redirecting the starch synthesis pathway toward fatty acid production. In this process, initially, microorganism is cultivated in nutrient-rich medium supplied with high concentration of CO_2 to improve the growth rate and biomass productivity after that transfer them to nutrient-deficient medium. Through this approach, the lipid content in microalgae was increased significantly, generally about 2- to 3.5-fold higher than normal conditions (Ho et al. 2012). However, a main challenge to using photosynthetic microorganism to produce biodiesel is the extraction of raw material and free fatty acids, for biodiesel production.

Current extraction processes are time-consuming and energy-intensive, requiring scientists to harvest the cells from the growth medium, dry them, and then extract the fatty acids with a variety of solvents. To avoid costly biomass recovery in photosynthetic microbial biofuel production, genetic engineering approach could be a promising technology to overcome the extraction hurdle. Recently, Liu et al. (2011) from Arizona State University (ASU) genetically discovered a solution. They redesigned the *Synechocystis* sp. PCC 6803 (SD100) by insertion of a vector with multiple gene substitutions of an acyl–acyl protein carrier thioesterase (*tesA*) gene (cause overproduction of fatty acids), to overproduce and secrete free fatty acids into the growth medium, which bypasses the need for deconstruction of biomass. The fatty acid secretion yield was increased to 197 ± 14 mg L^{-1} of culture in one improved strain at a cell density of 1.0×10^9 cells mL^{-1} by adding codon-optimized thioesterase genes and weakening polar cell wall layers. In this way, the fatty acids can be readily isolated from the growth medium, thereby averting many of the steps required to extract biofuels from the cyanobacterial cells. In addition, deletion of a penicillin-binding protein (PBP2) and introduction of an additional *tesA* gene of plant origin were also found to increase free fatty acid production in *Synechocystis* sp. PCC 6803. The free fatty acids accumulated in the medium as extracellular droplets with yields up to about 0.2 g L^{-1}, over 2–3 days of growth. Thus, bioprospecting of photosynthetic microorganism has the prospective to significantly impact on the total lipid content and consequently reduce the cost of biodiesel production. Researchers should look to explore the strains that are not only high lipid content but also have superior production rate and harvesting characteristics (Jones and Mayfield 2012). The high biomass and lipid productivity of photosynthetic microorganism warrants continuous investment in photosynthetic microorganism-based biodiesel research and makes them a potent candidate as a source of commercially viable biodiesel feedstock. With current growth systems, Rosenberg et al. (2011) estimated that the cost of producing biodiesel from algae in conjunction with an ethanol biorefinery would be in the range of 10–40 $ per gallon which is prohibitively expensive for biodiesel production. But it could be profitable with the incorporation of high-value algal coproducts.

5.4.1.2 Alcoholic Fuel

Alcohol-based fuels have been vital energy source since 1800s, used as fuel in three major approaches: combustion engines fuel (replacements for gasoline), fuel additives to octane boosting (antiknocking), and direct conversion of chemical energy to electrical energy (fuel cells) (Minteer 2006). The general chemical formula for alcohol fuel is $C_nH_{2n+1}OH$. Due to the presence of hydroxyl group with short carbon chain, it is a polar in nature and acts as a versatile solvent, i.e., miscible with water, organic solvents, and ionic compounds. The first four simplest aliphatic alcohols (methanol, ethanol, propanol, and butanol) are typically used as fuels because they can be easily synthesized chemically or biologically. Among all, ethanol is the most widely used liquid renewable fuel source, due to their lower toxicity and wide abundance.

Bioethanol Fuel

Apart from biodiesel, bioethanol is one of the most successful biofuels, as a clean and renewable combustible. Currently, it is produced by the fermentation and distillation of natural sugars derived mainly from sugarcane and various starchy crops: corn, potatoes, cassava, etc. These traditional crop sources are unable to meet the global demand of bioethanol production due to their crucial value in food or fuel dilemma. In recent years, due to increasing ethanol demand, ethanol production has been significantly boosted worldwide with increase in the quantity of CO_2 emission from ethanol fermentation plants. In 2008, 9,000 million gallons of ethanol were produced in the USA along with about 25.9 million metric tons of fuel-ethanol fermentation CO_2. In this context, photosynthetic microorganism holds significant potential to be a feedstock for ethanol production. They have higher photon conversion efficiency to harness sunlight via photosynthesis and fix the C_i from atmospheric or flue gas CO_2 emitted from petroleum-based power stations or other industrial sources which is then assimilated in the form of reserve food materials such as carbohydrate for bioethanol production. It is estimated that approximately 5000–15,000 gallons of ethanol/acre/year (4.67–14.0 L/m^2/year) can be produced by algae. This is 10–30 times higher than corn starch ethanol systems, which produces 400–500 gallons of ethanol/acre/year (0.37–0.46 L/m^2/year) (Cherry et al. 2008). Aquatic photosynthetic microorganisms are buoyant, evading the need for structural biopolymers such as hemicellulose and lignin. This in turn simplifies the process of bioethanol production by eliminating the chemical and enzymatic pretreatment steps to breakdown these biopolymers into fermentable sugars, as compared to other lignocellulosic terrestrial plants. Green microalgae and cyanobacteria typically accumulate starch or glycogen to a content of 10–50 % of their biomass, depending on the strain and growth conditions, and this polysaccharide is potentially useful as substrate for biofuel fermentation. Brown and Nobles (2007) patented a procedure to manufacture a potential feedstock for bioethanol fermentation, cellulose, photosynthetically in cyanobacteria. They expressed a cellulose synthesis gene cluster (*acs*ABC) of heterotrophic α-proteobacterium *Gluconacetobacter xylinus* ATCC 53582 in *Synechococcus elongatus* PCC 7942 and produced extracellular, non-crystalline cellulose in concentrations up to 0.22 g L^{-1}. Furthermore, Su et al. (2011) expressed same genes (*acs*AB) from heterotrophic bacterium *Acetobacter xylinum* in filamentous *Anabaena* sp. PCC 7120. Indeed, genome sequence analysis showed many cyanobacterial strains contain genes putatively involved in cellulose biosynthesis (Nobles et al. 2001). Thus, through the use of engineered photosynthetic microorganism, capacity of bioethanol production may significantly enhance and also able to eliminate food crops dependency. Among the most carbohydrate-rich microalgae, *Chlamydomonas reinharditii* (53–47 %), *C. vulgaris* (12–37 %),

Chlorella sp. (21–27 %), *Scenedesmus* sp. (13–20 %), *Spirogyra* sp. (33–64 %), *Porphyridium cruentum* (40–57 %), etc., have been shown to accumulate large amount of polysaccharides in their cell walls for bioethanol production. However, the higher starch content of microalgae can also be improved by optimizing the nitrogen or iron concentration during cultivation (Hirano et al. 1997). In addition, they can be also used for the production of both lipid-based biofuels and ethanol biofuels from the same biomass as a means to increase their overall economic value. Harun et al. (2010a) showed that the green algae *Chlorococum* sp. produces 60 % higher ethanol concentrations for samples that are pre-extracted for lipids versus those that remain as dried intact cells. In contrast to microalgae, naturally cyanobacteria produce very little amount of ethanol. However, cyanobacteria have some advantages over microalgae. Cyanobacterial cell wall contains a peptidoglycan layer, closely resembles that of Gram-positive bacteria. Thus, they are easily degradable by lysozyme and are less complex and less diverse than the cell walls of most microalgae. Moreover, cyanobacteria have glycogen as a storage carbohydrate, not reported in any eukaryotic microalgae. Glycogen may be preferred over starch as a fermentation feedstock. In the last decade, many researchers tried to engineer cyanobacteria to produce ethanol by insertion of pyruvate decarboxylase (*pdc*) and alcohol dehydrogenase II (*adh*) genes from the ethanol-producing *Zymomonas mobilis* into *S. elongatus* PCC 7942 and *Synechocystis* sp. PCC 6803 under the control of rbcLS promoter from RuBisCO (Deng and Coleman 1999; Dexter and Fu 2009). Inspite of, high activities of the *pdc* and *adh* genes, ethanol production was very low; 0.23 g L^{-1} (Deng and Coleman 1999) and 0.46 g L^{-1} (Dexter and Fu 2009). However, Gao et al. (2012) successfully constructed an efficient ethanol-producing mutant strain of *Synechocystis* sp. PCC6803 by introducing exogenous *pdc* from *Z. mobilis* and overexpressing endogenous *adh slr119*. The eventual ethanol concentration and productivity remarkably exceeded, 5.50 g L^{-1} and 212 mg L^{-1} day^{-1}, respectively, from previous reported. These results indicating that in cyanobacteria, careful metabolic engineering of the carbon allocating pathways is needed to obtain high levels of ethanol. However, two companies, Joule Unlimited company (Bedford, MA, USA) and Algenol (Bonita Springs, FL, USA) had claimed to have engineered photosynthetic microorganism to consume more than 90 % of the CO_2 and secrete ethanol directly in a continuous process. Recently, Algenol Biofuels Inc. developed a system which utilize sunlight trapping microalgal cells in the marginal or desert land and produce 6,000 gallons of ethanol per acre per year. This yield is far greater than the ethanol from corn which is only at a rate of 400 gallons of ethanol per acre per year (Dehring et al. 2012).

Biobutanol Fuel

Butanol from biomass is called biobutanol fuels, includes mainly isobutanol and n-butanol. Due to longer four carbons straight chained alcohol, it is nonpolar in nature, more similar to gasoline than ethanol. Because of the good properties of high heat value, high viscosity, low volatility, high hydrophobicity, and less corrosive, butanol has the potential to be a good fuel in the future. It can be mixed with ethanol, ether, and other organic solvent. Schwarz and Gapes (2006) demonstrated that n-butanol can be used either 100 % in unmodified 4-cycle ignition engines or blended up with diesel to at least 30 % in a diesel compression engine or blended up with kerosene to 20 % in a jet turbine engine. Isobutyraldehyde is a platform chemical for the synthesis of other chemicals and isobutanol which act as a potential biofuel for gasoline substitute. Isobutanol's potent biofuel properties include the following: anticorrosion property, more dense than gasoline so could be mix at any proportion with gasoline, and origin source is not connected with food supplies. Isobutyraldehyde produced industrially by the hydroformylation of propene, while Isobutanol is produced by the carbonylation of propylene. However, genetically engineered cyanobacteria are suited to biosynthesis of both isomers directly from CO_2 and increased productivity by overexpression of ribulose 1,5-bisphosphate carboxylase/oxygenase (RuBisCO). Recently, Atsumi et al. (2009) was successful in cloning the keto-acids pathway in *S. elongatus* PCC 7942 by introducing the oxoacid decarboxylase gene *kivD* from *Lactococcus lactis* under the control of the isopropyl-β-D-thiogalactoside inducible promoter P_{trc}. Increase in isobutyraldehyde production was obtained by expressing RuBisCO, which presumably increased the carbon flux through pyruvate. The maximum yield of isobutyraldehyde was 1.1 g/L over 8 days, higher rate than those reported for biofuels (ethanol, lipid, or hydrogen). Isobutanol production was obtained by expressing a suitable NADPH-dependent YqhD alcohol dehydrogenase along with kivD. Isobutanol yield of 0.45 g/L was achieved over 6 days. Although these results underscore the promise of direct bioconversion of CO_2 into fuels and chemicals, it is difficult to scale up the photosynthetic process.

A different approach was employed to manufacture n-butanol from CO_2 (Lan and Liao 2011), introduce a CoA-dependent pathway (transforms acetyl-CoA to 1-butanol) in *S. elongatus* sp. PCC 7942. However, 1-butanol production by this pathway is oxygen sensitive. Removal of oxygen is an important factor in the synthesis of 1-butanol in this organism. Therefore, optimal 1-butanol production (0.0145 g/L over 7 days) was obtained only under dark, anaerobic conditions where the cyanobacteria utilize their internal carbohydrate storage.

Bioisoprene Fuel

Isoprene (or 2-methyl-1, 3-butadiene) is a colorless, hydrophobic, and volatile organic liquid, derived in small amounts via plants such as oaks, poplars, eucalyptus, and legumes or industrially by thermal oil cracking. It can be used as renewable biofuel or as precursor feedstock for the production of synthetic rubber and other high-value compounds. The bioisoprene has higher energy content than other biofuels and has 80 % fewer greenhouse gases than petroleum-based fuels. For photosynthetic isoprene generation, cyanobacteria could act as both photocatalyst and producer of bioisoprene fuel from CO_2 and H_2O. Lindberg et al. (2010) engineered cyanobacterium *Synechocystis* sp. PCC 6803 genetically by introducing the isoprene synthase gene (*ispS*), operably linked to a PsbA2 promoter, from the plant *Pueraria montana*. However, the isoprene yield (about 50 μg isoprene per g dry cell weight per day) was very low. Therefore, severe research is required to enhance the yield through directing majority of the fixed carbon toward isoprene biosynthetic pathway.

5.4.1.3 Biogas

Biogas is a mixture of methane (55.0–65.0 %), carbon dioxide (30.0–45.0 %), traces of hydrogen sulfide, hydrogen, carbon monoxide, and water vapor, produced through the methanogenic decomposition of organic waste under anaerobic conditions (Costa and De Morais 2011). The process of anaerobic digestion occurs in three sequential stages: hydrolysis, fermentation, and methanogenesis (Molino et al. 2013). The hydrolysis of complex compounds is the breakdown of carbohydrates into soluble sugars. Afterward, the fermentation and methanogenesis carried out by bacteria which convert this biomass into alcohols, acetic acid, volatile fatty acids, and biogas. One of the valuable advantages of biogas is that after anaerobic digestion, the feedstock can be used as a biofertilizer, incinerated or used in animal feed. Recently, the potential of photosynthetic microorganisms as a source of biofuels and as a feedstock for anaerobic digestion has been subjected to intense academic and industrial research (Carver et al. 2011). In comparison with the traditional biogas feedstock plant biomass, photosynthetic microorganisms have many useful advantages such as grow in a liquid medium, less time and space required to grow, and lignin absent; thus, no pre-treatment is required for further processing and recapture of produced CO_2 in cell cultivation (Harun et al. 2010a, b). In photosynthetic microorganism-based biogas systems, solar energy is transformed into cellular energy using CO_2 from surroundings, which can subsequently be converted to the chemical energy of methane through the anaerobic fermentation of produced biomass by bacteria. The resultant methane can either be burned in a gas-turbine-generator system to produce electricity or, through catalytic processes, be converted to hydrocarbon fuel. Furthermore, to reach an economical balance, the conversion of algal biomass into biogas even recovers energy through the extraction of algal lipids that can be used for bioplastics or biodiesel production (Sialve et al. 2009). The effectiveness of biogas production has been shown to be species dependent (Jones and Mayfield 2012). The high content of protein in microalgal biomass can result in lowering the C/N ratio. Eventual increase in ammonia production inhibits anaerobic microorganisms. This limiting factor can be solved by codigestion with products containing a high C/N ratio (Brennan and Owende 2010). Yen and Brune (2007) achieved an increase in biogas production with the addition of high carbon content of waste paper residues in algal sludge feedstock. In this study, biogas production doubled (0.57 mL L^{-1} day^{-1}), using a waste/biomass ratio of 1:1, when compared to anaerobic digestion containing only algal biomass. Sodium ions can also be toxic for some anaerobic microorganisms, which can be minimized by preadapting the microorganisms that are going to be used (Brennan and Owende 2010). Recently, many researchers showed the potential of photosynthetic microorganism in biogas generation (Vanegas and Bartlett 2013; Prajapati et al. 2013). However, at present, the production of biogas from photosynthetic microorganisms is still limited may be due to infrastructure costs. Technically, to start a biogas plant need more land area and infrastructure to produce the same amount of energy as can be obtained for algae biodiesel.

5.4.1.4 Biohydrogen

Hydrogen emerged as one of the most promising clean, recyclable, and efficient alterative energy carriers of the future which has biggest energy content per weight of every recognized fuel. Since, it is oxidized to water, with no emission of carbon dioxide, not contribute to air pollution or global warming. One of the exciting features is that hydrogen can be directly use in fuel cells to generate electricity (Das and Veziroglu 2001), valuable for both transportation and domestic application. Unfortunately, nowadays, most of the hydrogen is produced by non-renewable production methods including electrolysis of water and thermocatalytic reformation of hydrogen-rich organic compounds. This requires a high-energy input which increases the cost of hydrogen production. Thus, there is an urgent need to develop novel economically feasible approaches. One of the approaches to make biohydrogen an economically viable alternative to other means of hydrogen production is biological production of hydrogen using microorganisms to convert solar energy into hydrogen gas. As shown in Fig. 5.4, biological hydrogen production process

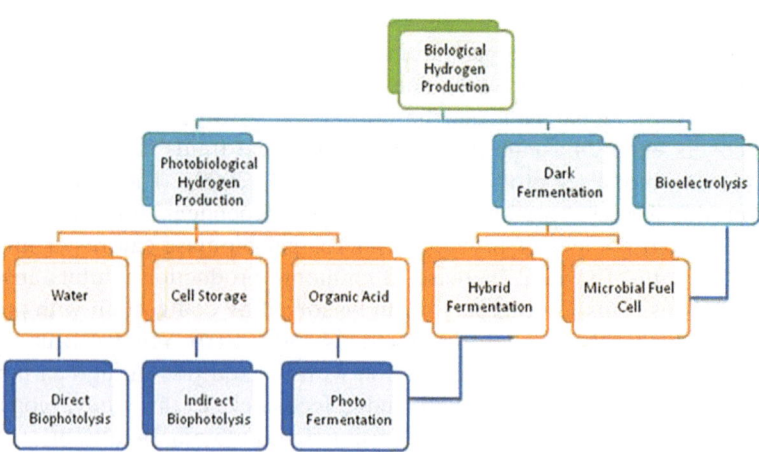

Fig. 5.4 Routes of biological hydrogen production process

can largely be separated into light/non-light energy dependent and further classified on their process feedstock sources.

Photosynthetic microorganisms have the genetic, metabolic, and enzymatic characteristics to produce hydrogen under anaerobic and limited aerobic conditions (Hwang et al. 2014). The unicellular green algal species *C. reinhardtii* is one of the well-known hydrogen-producing algae (Ghirardi et al. 2000; Melis and Happe 2001; Girbal et al. 2005). Hydrogenase activity has been reported in green algae *S. obliquus* (Florin et al. 2001; Girbal et al. 2005), in marine green algae *Tetraselmis subcordiformis* (Yan et al. 2011), *C. kessleri, C. vulgaris f. tertia, Ankistrodesmus braunii* (Kessler 1973), *Chlorococcum littorale* (Ueno et al. 1999; Guo et al. 2008), and *Playtmonas subcordiformis* (Guan et al. 2004), and in *Chlorella fusca* (Winkler et al. 2002). However, no hydrogenase activity was observed in *C. vulgaris, C. saccharophila, C. minutissima, C. vulgaris* (Kessler 1973), and *Duneliella salina* (Cao et al. 2001). Winkler et al. (2002) reported that *under sulpfr-depleted conditions,* enzyme activity of *S. obliquus* (150 nmol/μg Chl a h) is lower than that of *C. reindhartii* (200 nmol/μg Chl a h). Cyanobacterial hydrogen gas evolution involves heterocystous nitrogen-fixing cultures such as non-marine *Anabaena cylindrical* and *A. variabilis*, marine cyanobacter *Oscillatoria* sp. and *Calothrix* sp., and non-nitrogen-fixing organisms such as *Synechococcus* sp. and *Gloebacter* sp., and it was reported that *Anabaena* sp. have higher hydrogen evolution potential over the other cyanobacter species.

Mechanism of Biohydrogen Generation

As shown in Fig. 5.5, the photosynthetic microorganisms have ability to split water on photosystem II, i.e., produce electrons and liberate oxygen from one molecule of water, using sunlight as an energy source. Further, a light-dependent transfer of produced electrons to Fe–Fe-hydrogenase (H$_2$ase) in microalgae (Hwang et al. 2014) or to nitrogenase (N$_2$ase) in some cyanobacteria occurs, leading to the synthesis of molecular H$_2$ at photosystem I. All of the mechanisms discussed above are related to direct photolysis mechanisms. During this procedure, CO$_2$ is eventually fixed by the Calvin cycle into biosynthetic intermediates and storage compounds. In indirect photolysis, H$_2$ production continues either by oxidization of storage compounds through mitochondrial respiration or through dark fermentation of storage compound. The appeal of this system is that it uses water as a substrate and sunlight as an energy source and produced valuable biomass which can subsequently be exploited for other bioenergy as described above or again for hydrogen production via dark fermentation. Both of the precursors, water and sun light, are free and inexhaustibly available. The algal hydrogen production could be considered as an economical and sustainable method in terms of water utilization as a renewable resource and CO$_2$ consumption as one of the air pollutants. Thus, logically, this approach is tremendously capable to provide low-cost hydrogen production.

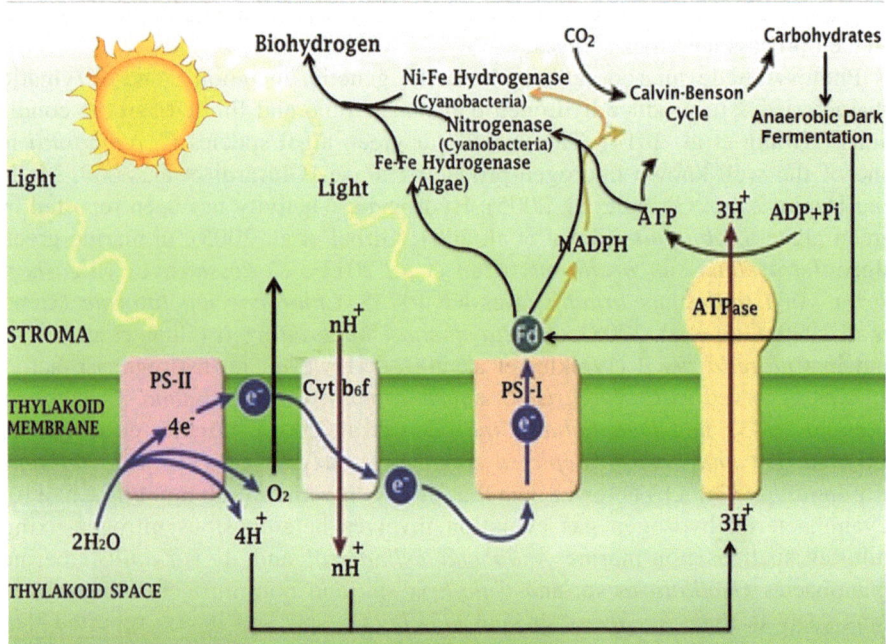

Fig. 5.5 Photosynthetic microorganism biohydrogen generation pathways

Strategies to Enhance the Biohydrogen Production

At present, the direct mechanism has many limitations as a tool of further research and for practical application. One of the obstacles to hydrogen production using photobiological systems is that key enzymes H_2ase and N_2ase are extremely sensitive to inactivation by oxygen evolved upon illumination by the water-oxidizing reactions of PSII (Lopes Pinto et al. 2002). Entailing the separation of two mutual incompatible O_2 and H_2 gases is a costly and technologically challenging feat. To overcome [Fe]-H_2ase issue, Melis and colleagues (2000) use sulfur deprivation strategy for lowering the partial pressure of oxygen and CO_2 fixation in green algae *C. reinhardtii*. In sulfur-deprived cultures, the starch content was significantly higher after H_2 production was over than before the cells were deprived of sulfur (Zhang et al. 2002). It is possible that H_2 production from sulfur-starved cultures will cause a similar increase followed by a decrease in lipid content. Lee and Greenbaum (2003) proposed another hypothesis that this oxygen sensitivity is a proton-gradient-related problem. Thus, it can be avoided by the genetic insertion of a hydrogenases promoter-programmed poly-peptide proton channel into the algal thylakoid membranes. Another strategy used by many researchers to overcome this issue is to implement a two-stage system. As the optimal operating conditions for the CO_2 uptake and H_2 production are different, two-stage system can be effectively employed to separate these two phases. In this system, photosynthetic microorganism cells fix CO_2 first to produce cellular substances starch (green algae) or glycogen (cyanobacteria) in aerobic conditions under light to stimulate oxygenic photosynthesis. This is followed by a second stage, where the cells are concentrated as needed, become anaerobic (oxygen limiting), which induces the H_2ase enzyme and start producing H_2 from their stored cellular substances in a dark fermentation. Both, green algae and a cyanobacterium, would carry out all these reactions and be reused through several such cycles of CO_2 fixation and H_2 evolution (with CO_2 recycled between stages). Yoon et al. (2002) carried out high-cell-density culture of *Anabaena variabilis* using repeated injections of CO_2 for the efficient production of H_2 in the second stage. Repeated injections of CO_2 were reported to increase cell density 3.7 g dry cell/l in 20 days (6.7 times higher than batch culture) during growth phase resulting in higher hydrogen production per gram cell in the later stage. Furthermore, to overcome the oxygen sensitivity of nitrogenase enzyme in cyanobacteria, addition of CO_2 and N_2 to the argon atmosphere is also reported for *Rahman* cultures during hydrogen evolution phase (Ferreira et al. 2012). One of the recent studies showed that rerouting the carbon pathways for carbohydrate catabolism and hydrogen production in cyanobacteria using metabolic engineering has also significant potential to enhance the yield of biohydrogen. Kumaraswamy et al. (2013) engineered the glycolysis pathway at the NAD^+-dependent glyceraldehyde-3-phosphate dehydrogenase (GAPDH-1) in *Synechococcus* sp. strain PCC 7002, resulting 2.3- to 3-fold increase in hydrogen than wild type. However, some studies reported that hydrogen production yield in cyanobacteria could also be redirect from the addition of minerals and nutrients. The presence of vanadium in Allen–Arnon medium induces the transcription of vanadium-based nitrogenase which is capable of evolving more hydrogen than molybdenum-based BG-11 and BG-11$_0$ media. Due to that, Allen–Arnon medium results 5.5-fold increase in hydrogen production by *A. variabilis* ATCC 29413 (Berberoaylu et al. 2008). In another study, addition of nitrate to cultures of *Synechococcus* sp. PCC 7002 acts as a metabolic switch which shifts the upper-glycolytic (EMP) pathway to the oxidative pentose phosphate (OPP) pathway and distribution of excreted products from predominantly lactate and H_2 to predominantly CO_2 and nitrite with increasing the total consumption of intracellular glycogen by threefold (McNeely et al. 2014). Thus, the reductant availability controls H_2 evolution in cyanobacteria. Another challenge in the area of biohydrogen is related to the light conversion efficiency of photosynthetic microorganisms. The potential light conversion efficiency to H_2 by photosynthetic microorganisms is theoretically about 10 %, due to 10 times slow electron transfer rate from PSII to PSI than photon capture rate by photosynthetic apparatus (Singh et al. 2013). This means that about 90 % of the photons being captured by the antenna systems are not being utilized (Mathews and Wang 2009). However, photosynthetic efficiency can be improved by truncating the chlorophyll antenna size of PS-II (Polle et al. 2002; Melis 2004; Kirst et al. 2012). Oey et al. (2013) showed the simultaneous knockdown of three LHC proteins (LHCMB1, 2 and 3) in the high H_2-producing *C. reinhardtii* mutant *Stm6Glc4* using an RNAi triple knockdown strategy, improved the light to H_2 (180 %) and biomass (165 %) conversion efficiencies at increased solar flux densities (450 instead of ~100 $\mu E\ m^{-2}\ s^{-1}$) and high cell densities. Recently, a light-harvesting complex protein LHCBM9 has been identified in microalga *C. reinhardtii*, serves an important protective function during stress conditions by promoting efficient light energy dissipation and stabilizing PSII super complexes (Grewe et al. 2014). Study showed that knockdown cell lines with 50–70 % reduced amounts of LHCBM9 showed reduced photosynthetic activity upon illumination and severe perturbation of hydrogen production activity. In contrast, presence of LHCBM9 in cells resulted in faster chlorophyll fluorescence decay and reduced production of singlet oxygen, indicating upgraded photoprotection.

In spite of the metabolic engineering, physiological and biochemical improvements in photosynthetic microorganism-based biohydrogen production, challenges are not entirely solved, and no commercial application has yet been announced. Biological production of hydrogen must be further investigated to make photosynthetic microorganism-based biofuel production energy and environmentally relevant.

5.4.2 Environmental Industry

Environmental industry is extremely broad which includes a wide range of goods and services that cut across many different industry sectors such as pollution control and prevention, environmental act regulations, remediation and management of toxic waste, sustainable energy, environmental protection and assessment (Rahman et al. 2011). However, in short, these consist of cleaner products and technologies, pollution management services, and resource management. Photosynthetic microorganisms may act as a powerful tool for the environmental industry, not only in control and management of environment, but also in revenue generation.

5.4.2.1 CO_2 Capture from Industry Sector Emissions

The global warming is an adverse incident attributed mainly to rise of GHGs in the atmosphere, particularly CO_2. This concern led to Kyoto Protocol promotion by United Nations with the objective of reducing GHGs by 5.2 % on the basis of the emissions in 1990 (Manne and Richels 2000). Many countries, Australia, Canada, USA, and Japan, had difficulty to accomplish their targets (Eboli and Davide 2012). Thus, there is an urgent need to develop novel CO_2 sequestration strategies which sustainably help to reduce the emissions from industry sector and accomplish the GHGs limit established by Kyoto Protocol. There are a wide variety of industrial activities that cause GHG emissions and many opportunities to reduce them. The strategies that have been intensively studied are included in the CCS methodologies. However, existing CCS methodologies have several technological, economical, and environmental issues, as well as safety problems, remain to be solved (Pires et al. 2012). As an alternative, the biological processes such as forestation, ocean fertilization, and photosynthetic microorganism cultivation can be applied to CO_2 capture. Among these, photosynthetic microorganisms have the capacity to fix CO_2 using solar energy with efficiency about 10–50 times greater than that of terrestrial plants (Chisti 2007). They are efficient enough to fix CO_2 in both atmosphere and industrial flue gases, which accumulate C_i in their cytoplasm to concentrations several orders of magnitude higher than that on the outside, phenomenon called CO_2-concentrating. Gas emissions from industries are collectively known as flue gases which contains not only CO_2, but also sulfur and nitrogen oxides (SO_x and NO_x). High SO_x and NO_x concentrations may inhibit the growth directly or by reducing the solution pH. Negoro et al. (1991) assessed the high CO_2-, SO_x-, and NO_x-level effects on growth of ten marine and halotolerant microalgal strains, with attention to the feasibility of making use of flue gas CO_2 as a carbon source. With 15 % of CO_2 concentration, the growth of green algae *Nannochloris* sp. and *Nannochloropsis* sp. was not affected by 50 mg L^{-1} of SO_2 gas. Nevertheless, as the SO_2 increased to 400 mg L^{-1}, the pH of the artificial seawater goes down and algae growth inhibited just after 20 h of cultivation. Both strains were also investigated with mixture of high CO_2 concentration and 300 mg L^{-1} of NO, resulting inhibition in growth of algae without any significant change of pH value. *Nannochloropsis* sp. did not grow, while *Nannochloris* sp. grew after a prolonged lag period. Despite these growth inhibitions, strains accumulated large amounts of crude lipids than normal conditions.

However, in contrast to the above results, recently, Lara-Gil et al. (2014) reported that SO_x and NO_x may serve as a sole nutrient source in a narrower range. They performed toxicity assays on *Desmodesmus abundans* UTEX-2976 and *Scenedesmus* sp. UTEX-1589, under 2 % (v/v) CO_2 and using nitrite, sulfite, or bisulfite salts. Interestingly, it observed that both nitrite (0–1,067 mg L^{-1} (w/v), NO_2^-) and sulfite (0–254 mg L^{-1} (w/v) SO_3^{2-}), not affect the growth at tested range. However, bisulfite was highly toxic just above 39 mg L^{-1}. Thus, direct utilization of flue gas might be possible with high flue gas tolerant strains. Where, NO_x and SO_x are beneficial nutritional components of the mitigation system. Further studies are needed to isolate and characterized the potent strains. As photosynthetic microorganisms consumes carbon dioxide, thus reducing harmful GHGs and earn carbon credit. The by-product left over after extracting the oil can be used in cattle feed, vitamins, pigments, cosmetics, etc., and algae can also be used to clean up waste water. Thus, photosynthetic microorganism-based CO_2 mitigation provides an opportunity for the companies to reduce CO_2 (and possibly earn carbon credits), while at the same time, other benefits also such as feedstock for biofuel, fertilizer, nutrient supplements, etc. Due to valuable end uses of produced biomass, CO_2 fixation by photosynthetic microorganisms could be economically more competitive than CCS methodologies.

5.4.2.2 Wastewater Treatment

Water is a vital commodity, and the access to hygienic useable water is a basic need and requirement (Singh et al. 2011). However, as the demand for clean water grows around the world, the need for wastewater treatment facilities capable of producing high-quality water becomes evident. With water scarcity and increasing incidents of pollutants' harmful effects in the world, photosynthetic microorganism-based wastewater treatment facilities could play a key role to provide a sustainable integrated water treatment solution. Algae cultivation does not have to affect the world's freshwater supply. In fact, unlike other biofuels, algae can be grown in anything from wastewater to saltwater. By using these two types of water, photosynthetic microorganisms put itself in a league of its own. Most traditional biofuel crops such as corn, soybeans, or jatropha, need freshwater for their growth. By utilizing wastewater, photosynthetic microorganisms actually use something that would essentially go to waste and help to filter out the pollutants in the process. Therefore, there are two benefits to using photosynthetic microorganisms: one, it helps to filter out pollutants before releasing the water into the environment, and two, the photosynthetic microorganism can be harvested and used for biofuel, fertilizer, nutrient supplements, etc. Table 5.6 shows the nutritional contents available in wastewater and their utilization in biochemical products of photosynthetic microorganism.

Nutrient removal coupled with metal detoxification could provide a sustainable, efficient, and cost-effective solution for the treatment of both municipal and industrial effluents. The photosynthetic microorganisms produce oxygen that allows aerobic bacteria to break down remaining contaminants in the water. Nutrients are removed from wastewater through the assimilation by growing algal cells. In addition to nutrient removal, chemical oxygen demand (COD) (Bich et al. 1999), total inorganic carbon (TOC) (Kim et al. 2010), and heavy metals can also be removed from wastewater through microalgal treatment (Mallick 2002). Photosynthetic microorganism has unique ability of a metabolic shift as a response to changes in the environmental conditions, and therefore, they are capable to assume many types of metabolisms such as autotrophic, heterotrophic, and mixotrophic.

Research pertaining to the use of photosynthetic microorganism for wastewater treatment began in the 1950 Oswald and Gotaas (1957) developed an advanced integrated high-rate algal ponds (HARPs) systems for combined wastewater treatment

Table 5.6 Wastewater nutrients and their utilization by photosynthetic microorganism

Nutritional content	Available forms	Function	Suitable content range
Carbon source	CO_2, CO_3^{2-}, HCO_3^-, etc.	Provide C to the hole cell, etc.	1–10 g/L
Nitrogen source	NO_3^-, Urea, AA, N_2, etc.	Provide N to the hole cell, etc.	10–2,000 mg/L
Phosphorus	Hydrophosphate, phosphate, etc.	Provide P to every reaction in cells, etc.	10–500 mg/L
Sulfur	Sulfate etc.	Provide S to proteins and reactions, etc.	1–200 mg/L
Inorganic salts	K, Ca, Na, Mg, etc.	Maintain cell structure and activity, etc.	0.1–100 mg/L
Trace elements	Fe, Zn, Mn, Pb, Cd, etc.	Be coenzyme factors, etc.	0.01–10 mg/L

and recovery of nutrients as algal biomass production. In the recent years, the use of photosynthetic microorganism to degrade or detoxify hazardous as bioremediants (Bhatnagar et al. 2010; Jiang et al. 2011; Marinho-Soriano et al. 2011; Rao et al. 2011; Min et al. 2011; Liu et al. 2012; Singh et al. 2014b, c) and biosorbents (Sivaprakash et al. 2010; Tuzen et al. 2009; Vogel et al. 2010; Rajfur et al. 2010; Mane and Bhosle 2012; Bermúdez et al. 2012; Singh et al. 2012) has been studied extensively. In order for photosynthetic microorganism-based wastewater treatment to be more sustainable, the photosynthetic microorganism production process must be coupled with other useful applications such as bioenergy production, carbon sequestration, and other value-added product production, as shown in Fig. 5.6.

Advances in Photosynthetic Microorganism-Based Wastewater Treatment System Coupled with CO_2 Capturing

Cunningham et al. (2010) predicted photosynthetic microorganism production process, coupled with a wastewater treatment plant and a power plant supplying flue gas, could cover its costs with a 10.47 % rate of return. In order to develop an economically feasible wastewater treatment system combined with carbon mitigation

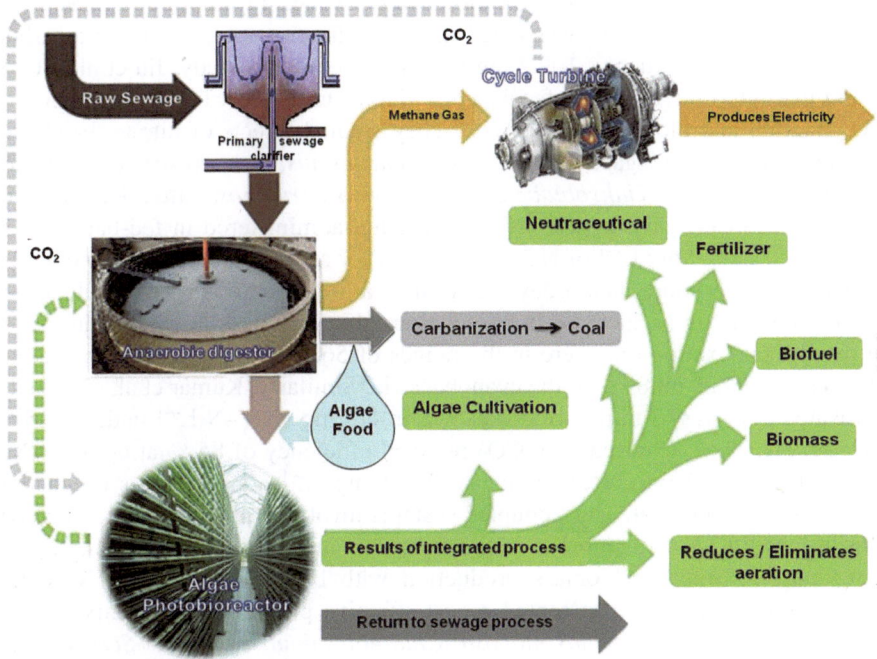

Fig. 5.6 Sustainable routes of photosynthetic microorganism-based wastewater treatment system coupled with CO_2 capturing

from flue gas, several researchers evaluated the many algal strains in the last decade. To develop an economically feasible system to remove ammonia from wastewater, Yun et al. (1997) cultivated *C. vulgaris* in wastewater and CO_2 from flue gas discharged from a steelmaking plant. CO_2 fixation and ammonia removal rates were estimated at 26 g m^{-3} h^{-1} and 0.92 g NH$_3$ m^{-3} h^{-1}, respectively, when wastewater was supplemented with external phosphate, without controlling the pH at 15 % (v/v) of CO_2. Many researchers reported that CO_2 fixation efficiency of photosynthetic microorganism may significantly affected by various operational cultivation conditions such as temperature, light duration, and nutrient limitations. Thus, they could be effectively applied as a tool to regulate the CO_2 fixation rate of photosynthetic microorganism, in order to prolong the active exponential phase of cell growth.

Chinnasamy et al. (2009) studied growth curve of *C. vulgaris* under varying CO_2 concentrations (ranging from 0.036 to 20 %) and temperature (30, 40, and 50 °C). They found that optimum temperature for biomass production (210 μg mL^{-1}) was 30 °C under 6 % elevated CO_2 level. Gomez-Villa et al. (2005) also reported different biomass production rates of 9 and 16 g L^{-1} d^{-1} in *S. obliquus* for winter and summer, respectively, which were equivalent to CO_2 fixation rates of 16 and 31 g CO_2 m^{-3} d^{-1}. Similarly, Jacob-Lopes et al. (2008) evaluated the influence of different photoperiod on the rates of CO_2 fixation by cyanobacteria *Aphanothece microscopica Nägeli* in refinery wastewater. Results indicated that the intermittent light regime had a strong impact on CO_2 fixation, resulting in a loss of 78 %. In optimum conditions, i.e., 11 klux, 35 °C, and 15 % CO_2, an increase of 58.1 % was noted in carbon fixation rate, providing a carbon fixation rate to the order of 109.2 mg L^{-1} h^{-1}. In another study, Jin et al. (2006) found that intermittent nitrate feeding is a viable strategy for the augmentation of CO_2 fixation and biomass productivity. They studied effects of nitrate feeding in two species of green algae, *Chlorella* and *Scenedesmus*, and two species of cyanobacteria, *Microcystis ichthyoblabe* and *Microcystis aeruginosa*, after adaptation to a 15 % (v/v) CO_2 environment. Nitrate feeding, administered in fed-batch mode to maintain 15–20 mg L^{-1} of NO$_3$–N, allowed for an extension of the exponential growth phase by more than 3 days, as well as a higher cell density, which subsequently resulted in an increase in photoautotrophic carbon fixation. The increases in the carbon fixation rate were in the ranges of 56.1–56.6 % for the green algae and between 68.2–68.8 % for the cyanobacteria. Similarly, Kumar et al. (2010a, b) cultivated *S. platensis* nitrate-rich wastewater 412 mg NO$_3^-$–N L^{-1} under 2–15 % CO_2 supply. They achieved high CO_2 removal efficiency of 85 % at an inlet CO_2 concentration of 2 %, corresponding to 2,131 mg L^{-1} algal biomass production and 68 % NO$_3$ removal. Thus, coupled systems involving photosynthetic microorganism-based CO_2 mitigation with wastewater treatment seems to be quite promising and for valuable biomass production with biological cleaning. Recently, Chinnasamy et al. (2010) patented a cost-effective process to grow mixotrophic algal strains (*Chlamydomonas globosa, Chlorella minutissima,* and *Scenedesmus bijuga*) in untreated wastewater of carpet industry, under the exposure of gas stream with 5–6 % of CO_2. However, economic feasibility of the process lies in the end uses of produced biomass.

5.4.3 Agriculture Industry

Agriculture is the science, art, or occupation concerned with domestication of plants and animals. It includes land cultivation, raising livestock of crops, breeding, feeding, and nurturing animals to provide food, wool, milk, and other products to sustain and enhance human life (Gupta et al. 2012). The agricultural industry is one of the oldest in the world, developed back around 10,000 BC as a key for expansion of ancient human civilizations (Goudie 2013). Due to significant advancement of science and engineering, over the past century, this sector has seen a lot of changes by improving soil fertilization, developing larger and hardier crops, and improving the nutrients in harvested food. Among all agricultural input, fertilizers are probably the most inevitable and basic requirement of modern intensive farming because of increased demand of nutrients of the high-yielding varieties (Lagreid et al. 1999). Of the fertilizers, physiologically nitrogen being an essential element deserves the vital position, considered as a yield-limiting factor. However, increased cost of the fertilizer is becoming an economic constrains for the small farmers of developing countries. Moreover, the continuous use of chemical fertilizers causes the ecological and biochemical imbalance in the agriculture (Yan et al. 2007).

In this context, photosynthetic microorganism could provide an exciting sustainable cost-effective solution of nitrogen. Historically, they are known to improve water-binding capacity and mineral composition of the soil (Barclay and Lewin 1985) and used for soil fertilization in coastal areas all over the world (Barsanti and Gualtieri 2014). Many free-living cyanobacterial strains fix atmospheric nitrogen and since they are photosynthetic, compete neither with crop plants nor with heterotrophic soil microflora for carbon and energy. Nitrogen-fixing ability has not only been shown by heterocystous cyanobacteria (*Nostoc, Azolla, Anabaena, Aulosira*, etc.) (Tiwari 1991; Abraham and Dhar 2010), but also by several non-heterocystous unicellular (*Synechococcus, Gloeocapsa, Aphanothece, Gloeothece*, etc.) (Rippka and Waterbury 1977) and filamentous (*Oscillatoria, Scytonema, Calothrix, Phormidium, Plectonema, Trichodesmium*, etc.) cyanobacteria (Herrero et al. 2001). The species with biofertilizer potential are the heterocystous, filamentous forms belonging to the order Nostocales and Stigonematales. Heterocyst is a unique double-layered organelle (Haselkorn 1978), which contains the enzyme nitrogenase to fix N_2 gas into ammonia (NH_3), nitrites (NO_2^-), or nitrates (NO_3^-), which can be easily absorbed by plants (atmospheric nitrogen is not bioavailable to plants), converted to protein and nucleic acids. In heterocystous, the nitrogenase activity and oxygenic photosynthesis are separated spatially and nitrogenase activity is usually light dependent. Many cyanobacterial strains *Nostoc, Anabaena, Tolypothrix, Aulosira, Cylindrospermum, Scytonema*, etc., are reported to found in rice fields and contribute significantly to their fertility (Vaishampayan et al. 2001; Gaydon et al. 2012). Besides nitrogen, many cyanobacteria *Tolypothrix, Scytonema, Hapalosiphon*, etc., have been also reported to solubilize rock phosphate to phosphorus (Prasanna and Kaushik 2006). Nevertheless, residual 80–90 % algal biomass which left after extraction of biomolecules for pharmaceutical or industrial use contains significant nutrient content (Prasanna and Kaushik 2006) can also be used directly as biofertilizer. Microalgae could also provide good animal overall nutrition due to its blend of proteins, carbohydrates, and vitamins. In aquaculture, they can be used for culturing several types of zooplankton that feed crustaceous and fish (Brown 2002). Egg-laying rates in hens are also reported to be increase by algae feed additives (Walker et al. 2012).

5.4.4 Food and Pharma Industry

As the world population is growing more than exponentially and expected to add another 1 billion in 7.2 billion of 2013 by 12 years (Camps and Engerman 2014). Thus, there is a growing unprecedented demand for additional ood sources, particularly food sources that are inexpensive to produce but nutritious (Gupta et al. 2013). To keep up with population and economic growth, food production should increase by 70 % by 2050 (Edmeades et al. 2010). Meat consumption is predicted to increase from 37 kg/person/year in 2000 to over 52 kg/person/year by 2050; if so, then 50 % of cereal production would go to animal feed (Nellemann 2009). Thus, new sustainable approaches are immediately needed to produce nutritious food for human and animals, and biofuels as well as to absorb CO$_2$, reduce the drain on freshwater agriculture and land. From thousands of year, photosynthetic microorganism biomass has been used as an individual food source and supplement with exceptionally high nutritional value (Skjanes et al. 2007). These organisms are widely known and consumed in certain regions, and numerous health benefits have been associated to their use. They are potentially a great source of natural compounds that could be used as ingredients for preparing functional foods. Many strains are rich in polyunsaturated fatty acids (PUFA) and their consumption helps to decrease in the incidence of cardiovascular diseases (Harwood and Guschina 2009). Microalgal carotenoids and PUFA possess anticancer features (Amaro et al. 2013). A high-value ketocarotenoid astaxanthin obtained from microalgal species, *Haematococcus pluvialis*, *Chlorella zofingiensis*, and *Chlorococcum* sp., (Yuan et al. 2011; Buono et al. 2014), possesses immunemodulatory action upon infection by *Helicobacter pylori* (Bhosale and Bernstein 2005). Acetylenic lipids from photosynthetic microorganism are known to have anticancer activity (Siddiq and Dembitsky 2008; Sheih et al. 2009). Furthermore, some strains, *Chaetoceros calcitrans*, *Skeletonema costatum*, *Phaeodactylum tricornutum*, *Chroomonas salina*, *Pavlova lutheri*, *Thalassiosira pseudonana*, *Isochrysis galbana*, *Prorocentrum micans*, etc., (Yongmanitchai and Ward 1991; Wang and Chai 1994) produce omega-3 fatty acids (ω-3 FA), eicosapentanoic acid (EPA), and decosahexaenoic acid (DHA) in superior quality compared to equivalent fatty acids from fish. Vazhappilly and Chen (1998) reported that *Monodus subterraneus* UTEX 151 (34.2 %), *Chlorella minutissima* UTEX 2341 (31.3 %), and *Phaeodactylum tricornutum* UTEX 642 (21.4 %) are EPA-rich strains, while *Crypthecodinium cohnii* UTEX L1649 (19.9 %), *Amphidinium carterae* UTEX

LB 1002 (17.0 %), and *Thraustochytrium aureum* ATCC 28211 (16.1 %) found to be rich in DHA proportion. High polyunsaturated DHA is known as an ingredient in infant formulas (Pyle et al. 2008) and anti-inflammatory properties (Orr and Bazinet 2008). While, EPA has superior lipid management properties, incredible anti-inflammatory effects (Heuvel et al. 2012), prevents and relieves painful symptoms of arthritis, lowering cholesterol and contributing to heart and cardiovascular health (Weaver and Holob 1987). EPA is also thought to have strong neuroprotective properties, positively affecting mental conditions such as schizophrenia and depression (Peet 2003).

Consequently, many photosynthetic microorganism cells such as *Spirulina, Chlorella, and Dunalliella* are capable of synthesizing all essential amino acids, carotenoids (especially provitamin A carotenoids), and other nutrients, such as vitamin B_{12} to humans and also animals (Kovač et al. 2013). Due to their high content and the amino acid pattern, they are good protein sources and commonly use as complementary dietary source, both directly as supplements in the form of tablets or powder, or as an additive in food and beverages (Tang and Suter 2011). Panjaitan et al. (2014) noted protein-rich *S. platensis* algae supplementation increases microbial protein production and feed intake and decreases retention time of digesta in the rumen of cattle. This microalga is also capable to produce high amounts of γ-linolenic acid (GLA). *Chlorella* has high levels of carotenoids such as lutein, zeaxanthin, and β-carotene is an effective dietary source of carotenoids for humans. Nagayama et al. (2014) reported *Chlorella* intake during pregnancy is effective in improving the carotenoid status of breast milk at early lactation. Breast milk carotenoids provide neonates with a source of vitamin A and potentially, oxidative stress protection, and other health benefits. *Chlorella* also yields an antibiotic known as chlorellin which affect Gram-positive and Gram-negative bacteria. *D. salina* is also able to produce 14% of β-carotene relative to its dry weight (Herrero et al. 2006). Algal biomass has also been shown to contain other antioxidants such as tocotrienols and tocopherols. These members of the vitamin E family are important antioxidants and have other health benefits such as protective effects against stroke-induced injuries, reversal of arterial blockage, growth inhibition of breast and prostate cancer cells, reduction in cholesterol levels, a reduced risk of type II diabetes, and protective effects against glaucomatous damage. In addition to these, photosynthetic microorganism biomass also considered as a functional food, illustrated by its immunological effects (Hayashi et al. 2008), reduction of maternal dioxin transfer (Nakano et al. 2005), promotion of intestinal bacteria (Pulz and Gross 2004), its effects on HIV-1 replication (Ayehunie et al. 1998), antioxidant activity (Miranda et al. 1998; Piñero Estrada et al. 2001), and several other health beneficial effects (Belay et al. 1993). Cyanobacterial strains such as *Arthospira plantesis* (*Spirulina*) produce sulfated polysaccharides named spirulan, endowed with several biological activities as antioxidant, anticoagulant, antiviral, and immune-stimulatory activity (Vonshak 1997). Another antiviral polysaccharide named nostoflan, isolated from a terrestrial cyanobacterium *Nostoc flagelliforme*, showed potent activity against herpes simplex viruses HSV-1, HSV-2, and other enveloped viruses (Kanekiyo et al. 2005).

Thus, over the last two decades or so, much research has been published on the food and pharmaceuticals applications. Nowadays, many researchers and companies are actively involved in large screening programs of bioprospecting with algal extracts to identify bioactive compounds of medical significance (Skjanes et al. 2007; León-Deniz et al. 2009; Mutanda et al. 2011). Their final objective is to develop new food products that can offer, besides the energetic and nutritional basic requirements, an additional benefit to human health. According to Pulz and Gross (2004), production of photosynthetic microorganism biomass for health, food, and animal feed/aquaculture is a fast-growing market, and market estimation suggests a retail value of 3,000–4,000 × 10^6 US dollar. However, more commercial potential for photosynthetic microorganisms still represents a largely untapped resource.

5.4.5 High-Value Compounds

The photosynthetic microorganism cells are immensely diversified and, due to physical and chemical properties of their polysaccharides, possess unique combination of various valuable functional properties such as non-toxicity, biodegradability, and renewable materials. The potential for storing the fixed CO_2 into large amount of algal biomass in combination with further industrial use is advantageous. The polysaccharides such as agar, alginates, and carrageenans have water-retention ability, film-forming capacity, and rheology, used in a wide variety of industrial applications, including food, antioxidants for preservation of cosmetics, sun protection, textiles, paper, and glycerol. In addition, other biochemical components of the cells include fatty acids, used in lipid-based cosmetics such as creams and lotions, and poly-β-hydroxybutyrate for production of plastics. Using photosynthetic microorganism to reinforce plastic materials provides satisfactory tensile properties, measured with up to 50 % dry weight of algal biomass, and these materials have a large variety of uses. Many composite materials have successfully been constructed from algae and polypropylene (Zhang et al. 2000a), polyethylene (Otsuki et al. 2004), and PVC (Zhang et al. 2000b). Species *Botrycoccus braunii* can store high amount of long-chain hydrocarbons (i.e., 25–31 carbon atoms), which can be utilized as substitutes of paraffinic and natural waxes (Metzger and Largeau 2005). Agar is also used in the preparation of ice cream and jellies, extracted from *Geliduim and Gracillaria* algae (Ito and Nagai 1998).

5.5 Challenges

The photosynthetic microorganisms are unique, valuable, and one of the most ecological versatile groups of microorganisms on our planet. It has been proposed to use as a tool to reduce the global warming for more than 50 years. Atmospheric

CO_2 could be a key resource for successful sustainable algal technology as carbon is the 36–65 % of dry matter of algal biomass (Chae et al. 2006; Sydney et al. 2010). However, this is not currently being done at industrial scale because of the associated high cost of biomass production. Despite its inherent potential in CO_2 fixation, many challenges have impeded the development of photosynthetic microorganism-based CO_2 mitigation technology to commercial viability that could allow for sustainable production and utilization as a bioenergy, feed, food, and fine chemicals. One of the challenge is L-CO_2 (380 ppm v CO_2) content of air (McGinn et al. 2011). Thus, diffusion of CO_2 from the atmosphere into a photosynthetic microorganism culture is not proficient to attain high biomass productivity. As an alternate solution, if C_i provide in the form of HCO_3^- salts or as compressed CO_2, it could increase to 41 % of total cost of the raw material (Molina Grima et al. 2003). Furthermore, it may also lead to noteworthy losses of CO_2 again into the atmosphere (Van Den Hende et al. 2012). In this context, use of flue gas containing CO_2 as an C_i source may be relatively a better option. However, besides CO_2, flue gas may also include up to 142 other compounds, e.g., carbon mono oxide, water, oxygen, halogen acids, nitrogen, nitrogen oxides, sulfur oxides, heavy metals, unburned carbohydrates, and particulate matters (Van Den Hende et al. 2012), some of which can be toxic for photosynthetic microorganism. Thus, in order to feed the flue gas to photosynthetic microorganism cultures, all flue gas compounds and their interactions with different strains need to be assessed for their tolerance. High CO_2 concentration in flue gas may decrease in pH level of culture medium, consequently reduces the activity of extracellular carbonic anhydrase which plays a major role in carbon concentration mechanism (Tang et al. 2011). Chiu et al. (2008) reported that *Chlorella* could able to tolerate CO_2 level up to 2 % only, more increase retard their growth rate. High cell density could be an answer to this challenge which minimizes the initial lag phase of photosynthetic microorganism and help to achieve high tolerance toward elevated CO_2 concentration. Overall, in order to obtain a successful sustainable photosynthetic microorganism-based CO_2 mitigation system to transform CO_2 into valuable products, some challenges should be addressed: (a) selection of potent strain with balance the needs of biofuel as well as extraction of value-added compounds, (b) optimization of reactor design to minimize the water evaporation and CO_2 diffusion losses, (c) development of continuous production systems to attain higher photosynthetic efficiencies, (d) incorporating flue gases and wastewater which are unsuitable in high concentration owing to the presence of poisonous compounds such as NO_x, SO_x, and heavy metals.

References

Abraham G, Dhar DW (2010) Induction of salt tolerance in *Azolla microphylla* Kaulf through modulation of antioxidant enzymes and ion transport. Protoplasma 245(1–4):105–111

Acton QA (2013) Aldehydes; Advances in research and application: 2013 Edition. ScholarlyEditions, Atlanta

Amaro HM, Barros R, Guedes A, Sousa-Pinto I, Malcata FX (2013) Microalgal compounds modulate carcinogenesis in the gastrointestinal tract. Trends Biotechnol 31(2):92–98

Aoyama K, Uemura I, Miyake J, Asada Y (1997) Fermentative metabolism to produce hydrogen gas and organic compounds in a cyanobacterium, *Spirulina platensis*. J Ferment Bioeng 83(1):17–20

Atsumi S, Higashide W, Liao JC (2009) Direct photosynthetic recycling of carbon dioxide to isobutyraldehyde. Nat Biotechnol 27(12):1177–1180

Ayehunie S, Belay A, Baba TW, Ruprecht RM (1998) Inhibition of HIV-1 replication by an aqueous extract of Spirulina platensis (Arthrospira platensis). JAIDS J Acquir Immune Defic Syndr 18(1):7–12

Barclay WR, Lewin RA (1985) Microalgal polysaccharide production for the conditioning of agricultural soils. Plant Soil 88(2):159–169

Barsanti L, Gualtieri P (2014) Algae: anatomy, biochemistry, and biotechnology, 2nd ed. Taylor & Francis, London

Belay A, Ota Y, Miyakawa K, Shimamatsu H (1993) Current knowledge on potential health benefits of Spirulina. J Appl Phycol 5(2):235–241

Berberoaylu H, Jay J, Pilon L (2008) Effect of nutrient media on photobiological hydrogen production by *Anabaena variabilis* ATCC 29413. Int J Hydrogen Energy 33(4):1172–1184

Bermúdez YG, Rico ILR, Guibal E, de Hoces MC, Martín-Lara MÁ (2012) Biosorption of hexavalent chromium from aqueous solution by Sargassum muticum brown alga. Application of statistical design for process optimization. Chem Eng J 183:68–76

Bhatnagar A, Bhatnagar M, Chinnasamy S, Das K (2010) *Chlorella minutissima*—a promising fuel alga for cultivation in municipal wastewaters. Appl Biochem Biotechnol 161:523–36

Bhosale P, Bernstein PS (2005) Microbial xanthophylls. Appl Microbiol Biotech 68(4):445–455

Bich NN, Yaziz MI, Bakti NAK (1999) Combination of *Chlorella vulgaris* and *Eichhrnia crassipes* for wastewater nitrogen removal. Water Res 33:2357–62

Brennan L, Owende P (2010) Biofuels from microalgae: a review of technologies for production, processing, and extractions of biofuels and co-products. Renew Sustain Energy Rev 14(2):557–577

Bridgwater AV (2003) Renewable fuels and chemicals by thermal processing of biomass. Chem Eng J 91(2):87–102

Brown MR (2002) Nutritional value and use of microalgae in aquaculture. Avances en Nutrición Acuícola VI Memorias del VI Simposium Internacional de Nutrición Acuícola 3:281–292

Brown RM, Nobles DR (2007) Expression of foreign cellulose synthase genes in photosynthetic prokaryotes (Cyanobacteria). US Patent

Buitenhuis ET, De Baar HJW, Veldhuis MJW (1999) Photosynthesis and calcification by *Emiliania huxleyi* (Prymnesiophyceae) as a function of inorganic carbon species. J Phycol 35(5):949–959

Buono S, Langellotti AL, Rinna F, Martello A, Fagliano V (2014) Functional Ingredients from Microalgae. Food Funct 5:1669–1685

Camps E, Engerman SL (2014) World population growth: the force of recent historical trends. J Interdiscipl Hist 44(4):509–526

Cao H, Zhang L, Melis A (2001) Bioenergetic and metabolic processes for the survival of sulfur-deprived *Dunaliella salina* (Chlorophyta). J Appl Phycol 13(1):25–34

Carver SM, Hulatt CJ, Thomas DN, Tuovinen OH (2011) Thermophilic, anaerobic co-digestion of microalgal biomass and cellulose for H_2 production. Biodegradation 22(4):805–814

Castenholz RW (1969) Thermophilic blue-green algae and the thermal environment. Bacteriol Rev 33(4):476

Chae SR, Hwang EJ, Shin HS (2006) Single cell protein production of *Euglena gracilis* and carbon dioxide fixation in an innovative photo-bioreactor. Bioresour Technol 97(2):322–329

Cherry J, Vlasenko E, Xu F (2008) U.S. Patent No. 7,354,743. U.S. Patent and Trademark Office, Washington, DC

Chiang C-L, Lee C-M, Chen P-C (2011) Utilization of the cyanobacteria Anabaena sp. CH1 in biological carbon dioxide mitigation processes. Bioresour Technol 102(9):5400–5405

Chinnasamy S, Bhatnagar A, Hunt RW, Claxton R, Marlowe M, Das KC (2010) Microalgae cultivation in a wastewater dominated by carpet mill effluents for biofuel applications. US Patent

Chinnasamy S, Ramakrishnan B, Bhatnagar A, Das KC (2009) Biomass production potential of a wastewater alga *Chlorella vulgaris* ARC 1 under elevated levels of CO_2 and temperature. Int J Mol Sci 10(2):518–532

Chisti Y (2007) Biodiesel from microalgae. Biotechnol Adv 25(3):294–306

Chisti Y (2008) Biodiesel from microalgae beats bioethanol. Trends Biotechnol 26(3):126–131

Chiu SY, Kao CY, Chen CH, Kuan TC, Ong SC, Lin CS (2008) Reduction of CO_2 by a high-density culture of *Chlorella* sp. in a semicontinuous photobioreactor. Bioresource technol 99(9):3389–3396

Costa JAV, De Morais MG (2011) The role of biochemical engineering in the production of biofuels from microalgae. Bioresour Technol 102(1):2–9

Costa JAV, Santana FB, Andrade MDR, Lima MB, Franck DT (2008) Microalga biomass and biomethane production in the south of Brazil. J Biotechnol 136:S430

Cuellar-Bermudez SP, Garcia-Perez JS, Rittmann BE, Parra-Saldivar R (2014) Photosynthetic bioenergy utilizing CO_2: an approach on flue gases utilization for third generation biofuels. J Clean Prod

Cunningham M, Heim C, Rauchenwald V (2010) Algae production in wastewater treatment: prospects for Ballen

Das D, Veziroğlu TN (2001) Hydrogen production by biological processes: a survey of literature. Int J Hydrogen Energy 26(1):13–28

Davies LL, Uchitel K, Ruple J (2013) Understanding barriers to commercial-scale carbon capture and sequestration in the United States: an empirical assessment. Energy Policy 59:745–761

Dehring U, Kramer D, Ziegler K (2012) Selection of ADH in genetically modified cyanobacteria for the production of ethanol. US Patent

Deng M-D, Coleman JR (1999) Ethanol synthesis by genetic engineering in cyanobacteria. Appl Environ Microbiol 65(2):523–528

Dexter J, Fu P (2009) Metabolic engineering of cyanobacteria for ethanol production. Energy Environ Sci 2(8):857–864

Diao Y-F, Zheng X-Y, He B-S, Chen C-H, Xu X-C (2004) Experimental study on capturing CO_2 greenhouse gas by ammonia scrubbing. Energy Convers Manag 45(13):2283–2296

Dlugokencky EJ, Crotwell A, Masarie K, White J, Lang P, Crotwell M (2013) NOAA measurements of long-lived greenhouse gases. Asia-Pacific GAW greenhouse gases: 6

Eboli F, Davide M (2012) The EU and Kyoto protocol: achievements and future challenges, review of environment, energy and economics (Re3). http://dx.doi.org/10.7711/feemre3.2012.12.003

Edmeades G, Fischer RA, Byerlee D (2010) Can we feed the world in 2050. In Proceedings of the New Zealand grassland association, vol 72. pp. 35–42

Ferreira AF, Marques AC, Batista AP, Marques PA, Gouveia L, Silva CM (2012) Biological hydrogen production by *Anabaena* sp.–Yield, energy and CO_2 analysis including fermentative biomass recovery. Int J Hydrogen Energy 37(1):179–190

Flint RW (2013) Basics of sustainable development. In: Practice of sustainable community development. Springer, New York, pp 25–54

Florin L, Tsokoglou A, Happe T (2001) A novel type of iron hydrogenase in the green alga *Scenedesmus obliquus* is linked to the photosynthetic electron transport Chain. J Biol Chem 276(9):6125–6132

Gao Z, Zhao H, Li Z, Tan X, Lu X (2012) Photosynthetic production of ethanol from carbon dioxide in genetically engineered cyanobacteria. Energy Environ Sci 5(12):9857–9865

Gaydon DS, Probert ME, Buresh RJ, Meinke H, Timsina J (2012) Modelling the role of algae in rice crop nutrition and soil organic carbon maintenance. Eur J Agron 39:35–43

Ghirardi ML, Zhang L, Lee JW, Flynn T, Seibert M, Greenbaum E, Melis A (2000) Microalgae: a green source of renewable H_2. Trends Biotechnol 18(12):506–511

Girbal L, von Abendroth G, Winkler M, Benton PMC, Meynial-Salles I, Croux C, Peters JW, Happe T, Soucaille P (2005) Homologous and heterologous overexpression in *Clostridium acetobutylicum* and characterization of purified clostridial and algal Fe-only hydrogenases with high specific activities. Appl Environ Microbiol 71(5):2777–2781

Gomez-Villa H, Voltolina D, Nieves M, Pina P (2005) Biomass production and nutrient budget in outdoor cultures of *Scenedesmus obliquus*(Chlorophyceae) in artificial wastewater, under the winter and summer conditions of Mazatlan, Sinaloa, Mexico. Vie et milieu 55(2):121–126

González-López CV, Acién Fernández FG, Fernández-Sevilla JM, Sánchez Fernández JF, Molina Grima E (2012) Development of a process for efficient use of CO_2 from flue gases in the production of photosynthetic microorganisms. Biotechnol Bioeng 109(7):1637–1650

Gore A (2006) An inconvenient truth: the planetary emergency of global warming and what we can do about it. Rodale, New York

Goswami RCD, Kalita MC (2012) Microalgal resources in Chandrapur area, North–East, Assam, India: a perspective for industrial refinement system and a boon for alternative energy generation and mitigation of green house gases. Arch Appl Sci Res 4(2):795–799

Goudie AS (2013) The human impact on the natural environment: past, present, and future, 7th edn. Wiley, New York

Gouveia L, Oliveira AC (2009) Microalgae as a raw material for biofuels production. J Ind Microbiol Biotechnol 36(2):269–274

Grewe S, Ballottari M, Alcocer M, Dae Andrea C, Blifernez-Klassen O, Hankamer B, Mussgnug JH, Bassi R, Kruse O (2014) Light-harvesting complex protein LHCBM9 is critical for photosystem II activity and hydrogen production in *Chlamydomonas reinhardtii*. Plant Cell Online 26:1598–1611. doi:10.1105/tpc.114.124198

Groom MJ, Gray EM, Townsend PA (2008) Biofuels and biodiversity: principles for creating better policies for biofuel production. Conserv Biol 22(3):602–609

Guan Y, Zhang W, Deng M, Jin M, Yu X (2004) Significant enhancement of photobiological H_2 evolution by carbonylcyanide m-chlorophenylhydrazone in the marine green alga *Platymonas subcordiformis*. Biotechnol Lett 26(13):1031–1035

Gupta E, Sinha J, Dubey RP (2012) Utilization of dehydrated herbs in the formulation of value added snack "rice flakes mix". J Food Process Technol S1-002. doi:10.4172/2157-7110

Gupta E, Purwar S, Sundaram S, Rai GK (2013) Nutritional and therapeutic values of Stevia rebaudiana: a review. J Med Plants Res 7(46):3343–3353

Guo Z, Chen Z, Zhang W, Yu X, Jin M (2008) Improved hydrogen photoproduction regulated by carbonylcyanide m-chlorophenylhrazone from marine green alga *Platymonas subcordiformis* grown in CO_2-supplemented air bubble column bioreactor. Biotechnol Lett 30(5):877–883

Harun R, Danquah MK, Forde GM (2010a) Microalgal biomass as a fermentation feedstock for bioethanol production. J Chem Technol Biotechnol 85(2):199–203

Harun R, Singh M, Forde GM, Danquah MK (2010b) Bioprocess engineering of microalgae to produce a variety of consumer products. Renew Sustain Energy Rev 14(3):1037–1047

Harwood JL, Guschina IA (2009) The versatility of algae and their lipid metabolism. Biochimie 91(6):679–684

Haselkorn R (1978) Heterocysts. Ann Rev Plant Physiol 29(1):319–344

Hayashi K, Nakano T, Hashimoto M, Kanekiyo K, Hayashi T (2008) Defensive effects of a fucoidan from brown alga *Undaria pinnatifida* against herpes simplex virus infection. Inter Immunopharmacol 8(1):109–116

Herrero A, Muro-Pastor AM, Flores E (2001) Nitrogen control in cyanobacteria. J Bacteriol 183(2):411–425

Herrero M, Cifuentes A, Ibanez E (2006) Sub-and super-critical fluid extraction of functional ingredients from different natural sources: Plants, food-by-products, algae and microalgae: a review. Food Chem 98(1):136–148

Heuvel JPV, Bhatia SK, Belcher LA, Thompson JT, Gillies PJ (2012) Regulation of inflammatory and lipid metabolism genes by eicosapentaenoic acid-rich oil. FASEB J 26(266):266

Hirano A, Ueda R, Hirayama S, Ogushi Y (1997) CO_2 fixation and ethanol production with microalgal photosynthesis and intracellular anaerobic fermentation. Energy 22(2):137–142

Hirata S, Hayashitani M, Taya M, Tone S (1996) Carbon dioxide fixation in batch culture of *Chlorella* sp. using a photobioreactor with a sunlight-collection device. J Ferment Bioeng 81(5):470–472

Ho S-H, Chen C-Y, Chang J-S (2012) Effect of light intensity and nitrogen starvation on CO_2 fixation and lipid/carbohydrate production of an indigenous microalga *Scenedesmus obliquus* CNW-N. Bioresour Technol 113:244–252

Hockstad L, Cook B (2012) Inventory of US greenhouse gas emissions and sinks: 1990–2010 (EPA 430-R-12-001). US Environmental Protection Agency, Washington, DC

Hwang J-H, A H-C, Choi J-A, Abou-Shanab RAI, Dempsey BA, Regan JM, Kim JR, Song H, Nam I-H, Kim S-N (2014) Photoautotrophic hydrogen production by eukaryotic microalgae under aerobic conditions. Nature Commun 5. doi:10.1038/ncomms4234

Ito E, Nagai H (1998) Morphological observations of diarrhea in mice caused by aplysiatoxin, the causative agent of the red alga *Gracilaria coronopifolia* poisoning in Hawaii. Toxicon 36(12):1913–1920

Iyovo GD, Du G, Chen J (2010) Sustainable bioenergy bioprocessing: biomethane production, digestate as biofertilizer and as supplemental feed in algae cultivation to promote algae biofuel commercialization. J Microb Bioche Technol 2(4):100–106

Jacob-Lopes E, Cacia Ferreira Lacerda LM, Franco TT (2008) Biomass production and carbon dioxide fixation by *Aphanothece microscopica Nageli* in a bubble column photobioreactor. Biochem Eng J 40(1):27–34

Jaggi B, Freedman M, Martin C (2011) Global warming, Kyoto Protocol, and the need for corporate pollution disclosures in India: a case study. Int J Bus Humanit Technol 1(3):60–67

Jiang L, Luo S, Fan X, Yang Z, Guo R (2011) Biomass and lipid production of marine microalgae using municipal wastewater and high concentration of CO_2. Appl Energy 88(10):3336–3341

Jin H-F, Lim B-R, Lee K (2006) Influence of nitrate feeding on carbon dioxide fixation by microalgae. J Environ Sci Health Part A 41(12):2813–2824

John RP, Anisha GS, Nampoothiri KM, Pandey A (2011) Micro and macroalgal biomass: a renewable source for bioethanol. Bioresour Technol 102(1):186–193

Jones CS, Mayfield SP (2012) Algae biofuels: versatility for the future of bioenergy. Curr Opin Biotechnol 23(3):346–351

Kanekiyo K, Lee JB, Hayashi K, Takenaka H, Hayakawa Y, Endo S, Hayashi T (2005) Isolation of an antiviral polysaccharide, nostoflan, from a terrestrial cyanobacterium, Nostoc flagelliforme. J Nat Prod 68(7):1037–1041

Kessler E (1973) Effect of anaerobiosis on photosynthetic reactions and nitrogen metabolism of algae with and without hydrogenase. Archive for Mikrobiologie 93(2):91–100

Kim J, Lingaraju BP, Rheaume RL, Joo-youp S, Kaniz F (2010) Removal of ammonia from wastewater effluent by *Chlorella vulgaris*. Tsinghua Sci Technol 15(4):391–396

Kirst H, Garcia-Cerdan JG, Zurbriggen A, Ruehle T, Melis A (2012) Truncated photosystem chlorophyll antenna size in the green microalga *Chlamydomonas reinhardtii* upon deletion of the TLA3-CpSRP43 gene. Plant Physiol 160(4):2251–2260

Kovač DJ, Simeunović JB, Babić OB, Mišan AČ, Milovanović IL (2013) Algae in food and feed. Food Feed Res 40:21–32

Kumar A, Ergas S, Yuan X, Sahu A, Zhang Q, Dewulf J, Malcata FX, Van Langenhove H (2010a) Enhanced CO_2 fixation and biofuel production via microalgae: recent developments and future directions. Trends Biotechnol 28(7):371–380

Kumar A, Yuan X, Sahu AK, Dewulf J, Ergas SJ, Van Langenhove H (2010b) A hollow fiber membrane photo-bioreactor for CO_2 sequestration from combustion gas coupled with wastewater treatment: a process engineering approach. J Chem Technol Biotechnol 85(3):387–394

Kumaraswamy GK, Guerra T, Qian X, Zhang S, Bryant DA, Dismukes GC (2013) Reprogramming the glycolytic pathway for increased hydrogen production in cyanobacteria: metabolic engineering of NAD^+-dependent GAPDH. Energy Environ Sci 6(12):3722–3731

Lagreid M, Bockman OC, Kaarstad O (1999) Agriculture, fertilizers and the environment. CABI publishing

Lan EI, Liao JC (2011) Metabolic engineering of cyanobacteria for 1-butanol production from carbon dioxide. Metab Eng 13(4):353–363

Lara-Gil JA, Álvarez MM, Pacheco A (2014) Toxicity of flue gas components from cement plants in microalgae CO_2 mitigation systems. J Appl phycol 1–12

Lee JW, Greenbaum E (2003) A new oxygen sensitivity and its potential application in photosynthetic H_2 production. Biotech Fuels Chemicals 47(2):303–313

León-Deniz LV, Dumonteil E, Moo-Puc R, Freile-Pelegrin Y (2009) Antitrypanosomal in vitro activity of tropical marine algae extracts. Pharm Biol 47(9):864–871

Li Q, Du W, Liu D (2008) Perspectives of microbial oils for biodiesel production. Appl Microbiol Biotechnol 80(5):749–756

Leung DYC, Wu X, Leung MKH (2010) A review on biodiesel production using catalyzed transesterification. Appl Energy 87(4):1083–1095

Lin K (2014) Approximate dynamic programming applied to biofuel markets in the presence of renewable fuel standards. Princeton University, Princeton

Lindberg P, Park S, Melis A (2010) Engineering a platform for photosynthetic isoprene production in cyanobacteria, using *Synechocystis* as the model organism. Metab Eng 12(1):70–79

Liu X, Sheng J, Curtiss Iii R (2011) Fatty acid production in genetically modified cyanobacteria. Proc Natl Acad Sci 108(17):6899–6904

Liu C-H, Chang C-Y, Cheng C-L, Lee D-J, Chang J-S (2012) Fermentative hydrogen production by *Clostridium butyricum* CGS5 using carbohydrate-rich microalgal biomass as feedstock. Int J Hydrogen Energy 37(20):15458–15464

Lopes Pinto FA, Troshina O, Lindblad P (2002) A brief look at three decades of research on cyanobacterial hydrogen evolution. Int J Hydrogen Energy 27(11):1209–1215

Mallick N (2002) Biotechnological potential of immobilized algae for wastewater N, P and metal removal: a review. Biometals 15(4):377–390

Mane PC, Bhosle AB (2012) Bioremoval of some metals by living algae *Spirogyra* sp. and *Spirullina* sp. from aqueous solution. Inter J Env Res 6(2):571–576

Manne AS, Richels RG (2000) The Kyoto Protocol: a cost-effective strategy for meeting environmental objectives? In: Efficiency and equity of climate change policy. Springer, New York, pp 43–61

Marinho-Soriano E, Azevedo CAA, Trigueiro TG, Pereira DC, Carneiro MAA, Camara MR (2011) Bioremediation of aquaculture wastewater using macroalgae and Artemia. Inter Biodeterior Biodegradation 65(1):253–257

Mata TM, Martins AA, Caetano NS (2010) Microalgae for biodiesel production and other applications: a review. Renew Sustain Energy Rev 14(1):217–232

Martinot E (2013) Renewable energy policy network for the 21st century (REN21's): global status report. Worldwatch Institute, Paris

Mathews J, Wang G (2009) Metabolic pathway engineering for enhanced biohydrogen production. Int J Hydrogen Energy 34(17):7404–7416

McGinn PJ, Dickinson KE, Bhatti S, Frigon JC, Guiot SR, O'Leary SJ (2011) Integration of microalgae cultivation with industrial waste remediation for biofuel and bioenergy production: opportunities and limitations. Photosynth Res 109(1–3):231–247

McNeely K, Kumaraswamy GK, Guerra T, Bennette N, Ananyev G, Dismukes GC (2014) Metabolic switching of central carbon metabolism in response to nitrate: application to autofermentative hydrogen production in cyanobacteria. J Biotechnol

Melis A, Happe T (2001) Hydrogen production. Green algae as a source of energy. Plant Physiol 127(3):740–748

Melis T (2004) Maximizing photosynthetic efficiencies and hydrogen production in microalgal cultures. In: Proceedings of 2004 DOE hydrogen program review, Philadelphia

Melis A, Zhang L, Forestier M, Ghirardi ML, Seibert M (2000) Sustained photobiological hydrogen gas production upon reversible inactivation of oxygen evolution in the green alga *Chlamydomonas reinhardtii*. Plant Physiol 122(1):127–136

Metzger P, Largeau C (2005) *Botryococcus braunii*: a rich source for hydrocarbons and related ether lipids. Appl Microbiol Biotechnol 66(5):486–496

Miao X, Wu Q (2004) High yield bio-oil production from fast pyrolysis by metabolic controlling of *Chlorella prototothecoides*. J Biotechnol 110(1):85–93

Miller G, Spoolman S (2008) Living in the environment: principles, connections, and solutions. Cengage Learning, India

Min M, Wang L, Li Y, Mohr M, Hu B, Zhou W, Chen P, Ruan R (2011) Cultivating *Chlorella* sp. in a pilot-scale photobioreactor using centrate wastewater for microalgae biomass production and wastewater nutrient removal. Appl Biochem Biotech 165(1):123–137

Minteer S (2006) Alcoholic Fuels. CRC Press, Boca Raton, FL; Taylor & Francis, London

Miranda MS, Cintra RG, Barros SBM, Mancini-Filho J (1998) Antioxidant activity of the microalga Spirulina maxima. Braz J Med Biol Res 31(8):1075–1079

Molina Grima E, Belarbi EH, Aciacn Fernaịndez FG, Robles Medina A, Chisti Y (2003) Recovery of microalgal biomass and metabolites: process options and economics. Biotechnol Adv 20(7):491–515

Molino A, Nanna F, Ding Y, Bikson B, Braccio G (2013) Biomethane production by anaerobic digestion of organic waste. Fuel 103:1003–1009

Mollenkopf H, Fozard JL (2003) Technology and the good life: Challenges for current and future generations of aging people. Ann Rev Gerontol Geriatr 23:250–279

Mostafa SSM, El-Gendy NS (2013) Evaluation of fuel properties for microalgae *Spirulina platensis* bio-diesel and its blends with Egyptian petro-diesel. Arab J Chem

Muradyan EA, Klyachko-Gurvich GL, Tsoglin LN, Sergeyenko TV, Pronina NA (2004) Changes in lipid metabolism during adaptation of the *Dunaliella salina* photosynthetic apparatus to high CO_2 concentration. Russ J Plant Physiol 51(1):53–62

Mussatto SI, Dragone G, Guimaraes PMR, Silva JPA, Carneiro LM, Roberto IC, Vicente A, Domingues L, Teixeira JA (2010) Technological trends, global market, and challenges of bio-ethanol production. Biotechnol Adv 28(6):817–830

Mutanda T, Ramesh D, Karthikeyan S, Kumari S, Anandraj A, Bux F (2011) Bioprospecting for hyper-lipid producing microalgal strains for sustainable biofuel production. Bioresour Technol 102(1):57–70

Nagayama J, Noda K, Uchikawa T, Maruyama I, Shimomura H, Miyahara M (2014) Effect of maternal *Chlorella* supplementation on carotenoid concentration in breast milk at early lactation. Int J Food Sci Nutr 0:1–4

Nakano S, Noguchi T, Takekoshi H, Suzuki G, Nakano M (2005) Maternal-fetal distribution and transfer of dioxins in pregnant women in Japan, and attempts to reduce maternal transfer with *Chlorella* (*Chlorella pyrenoidosa*) supplements. Chemosphere 61(9):1244–1255

Nellemann C (ed) (2009) The environmental food crisis: the environment's role in averting future food crises: a UNEP rapid response assessment. UNEP/Earthprint

Negoro M, Shioji N, Miyamoto K, Micira Y (1991) Growth of microalgae in high CO_2 gas and effects of SO_x and NO_x. Appl Biochem Biotechnol 28(1):877–886

National Aeronautics and Space Administration (2014) Global land-ocean temperature index in 0.01 °C: Goddard Institute for Space Studies. http://data.giss.nasa.gov/gistemp/tabledata_v3/GLB.Ts+dSST.txt

Nobles DR, Romanovicz DK, Brown RM (2001) Cellulose in cyanobacteria. Origin of vascular plant cellulose synthase? Plant physiology 127(2):529–542

Oey M, Ross IL, Stephens E, Steinbeck J, Wolf J, Radzun KA, Kagler J, Ringsmuth AK, Kruse O, Hankamer B (2013) RNAi knock-down of LHCBM1, 2 and 3 increases photosynthetic H_2 production efficiency of the green alga *Chlamydomonas reinhardtii*. PLoS ONE 8(4):e61375

Olaizola M (2003a) Commercial development of microalgal biotechnology: from the test tube to the marketplace. Biomol Eng 20(4):459–466

Olaizola M (2003b) Microalgal removal of CO_2 from flue gases: changes in medium pH and flue gas composition do not appear to affect the photochemical yield of microalgal cultures. Biotechnol Bioprocess Eng 8(6):360–367

Olivier JGI, Peters JAHW, Janssens-Maenhout G (2012) Trends in global CO_2 emissions 2012 report. PBL Netherlands Environmental Assessment Agency

Ono E, Cuello JL (2007) Carbon dioxide mitigation using thermophilic cyanobacteria. Biosyst Eng 96(1):129–134

Orr SK, Bazinet RP (2008) The emerging role of docosahexaenoic acid in neuroinflammation. Curr Opin Investig Drugs (London, England: 2000) 9(7):735–743

Oswald WJ, Gotaas HB (1957) Photosynthesis in sewage treatment. Trans Am Soc Civil Eng 122:73–105

Ota M, Kato Y, Watanabe H, Watanabe M, Sato Y, Smith RL Jr, Inomata H (2009) Fatty acid production from a highly CO_2 tolerant alga, *Chlorocuccum littorale*, in the presence of inorganic carbon and nitrate. Bioresour Technol 100(21):5237–5242

Otsuki T, Zhang F, Kabeya H, Hirotsu T (2004) Synthesis and tensile properties of a novel composite of Chlorella and polyethylene. J Appl Polym Sci 92(2):812–816

Panjaitan T, Quigley S, McLennan S, Swain T, Poppi D (2014) Spirulina (*Spirulina platensis*) algae supplementation increases microbial protein production and feed intake and decreases retention time of digesta in the rumen of cattle. Anim Prod Sci

Pearce DW (1992) The secondary benefits of greenhouse gas control. Centre for Social and Economic Research on the Global Environment

Peet M (2003) Eicosapentaenoic acid in the treatment of schizophrenia and depression: rationale and preliminary double-blind clinical trial results. Prostaglandins Leukot Essent Fatty Acids 69(6):477–485

Piñero Estrada JE, Bermejo Bescos P, Villar del Fresno AM (2001) Antioxidant activity of different fractions ofi *Spirulina platensis* protean extract. Il Farmaco 56(5):497–500

Pires JCM, Alvim-Ferraz MCM, Martins FG, Simaцes M (2012) Carbon dioxide capture from flue gases using microalgae: engineering aspects and biorefinery concept. Renew Sustain Energy Rev 16(5):3043–3053

Plaza MG, Pevida C, Arenillas A, Rubiera F, Pis JJ (2007) CO_2 capture by adsorption with nitrogen enriched carbons. Fuel 86(14):2204–2212

Polle JEW, Kanakagiri S, Jin E, Masuda T, Melis A (2002) Truncated chlorophyll antenna size of the photosystems: A practical method to improve microalgal productivity and hydrogen production in mass culture. Int J Hydrogen Energy 27(11):1257–1264

Powell EE, Hill GA (2009) Economic assessment of an integrated bioethanol–biodiesel–microbial fuel cell facility utilizing yeast and photosynthetic algae. Chem Eng Res Des 87(9):1340–1348

Prajapati SK, Kaushik P, Malik A, Vijay VK (2013) Phycoremediation and biogas potential of native algal isolates from soil and wastewater. Bioresource Technol 135:232–238

Prasanna R, Kaushik BD (2006) Cyanobacteria in soil health and sustainable agriculture. Health and Environment 3:91–105

Pronina NA, Rogova NB, Furnadzhieva S, Klyachko-Gurvich GL (1998) Effect of CO_2 concentration on the fatty acid composition of lipids in *Chlamydomonas reinhardtii* Cia-3, a mutant deficient in CO_2-concentrating mechanism. Russ J Plant Physiol 45(4):447–455

Pulz O, Gross W (2004) Valuable products from biotechnology of microalgae. Appl Microbiol Biotechnol 65(6):635–648

Pyle DJ, Garcia RA, Wen Z (2008) Producing docosahexaenoic acid (DHA)-rich algae from biodiesel-derived crude glycerol: effects of impurities on DHA production and algal biomass composition. J Agric Food Chem 56(11):3933–3939

Rahman MA, Soumya KK, Tripathi A, Sundaram S, Singh S, Gupta A (2011) Evaluation and sensitivity of cyanobacteria, *Nostoc muscorum* and Synechococcus PCC 7942 for heavy metals stress—a step toward biosensor. Toxicol Env Chem 93(10):1982–1990

Rajfur M, Kłos A, Wacławek M (2010) Sorption properties of algae *Spirogyra* sp. and their use for determination of heavy metal ions concentrations in surface water. Bioelectrochemistry 80(1):81–86

Rippka R, Waterbury JB (1977) The synthesis of nitrogenase by non-heterocystous cyanobacteria. FEMS Microbiol Lett 2(2):83–86

Ramaraj R, Tsai DD-W, Chen PH (2010) Freshwater microalgae niche of air carbon dioxide mitigation. Ecol Eng 68:47–52

Ramaraj R, Tsai DD-W, Chen PH (2014) Freshwater microalgae niche of air carbon dioxide mitigation. Ecol Eng 68:47–52

Rao HP, Ranjith Kumar R, Raghavan BG, Subramanian VV, Sivasubramanian V (2011) Application of phycoremediation technology in the treatment of wastewater from a leather-processing chemical manufacturing facility. Water SA 37:07–14

Rosenberg JN, Mathias A, Korth K, Betenbaugh MJ, Oyler GA (2011) Microalgal biomass production and carbon dioxide sequestration from an integrated ethanol biorefinery in Iowa: a technical appraisal and economic feasibility evaluation. Biomass Bioenergy 35(9):3865–3876

Sander K, Murthy GS (2010) Life cycle analysis of algae biodiesel. Int J Life Cycle Assess 15(7):704–714

Schwarz WH, Gapes JR (2006) Butanol-rediscovering a renewable fuel. BioWorld. Europe 01-2006:16–19

Shakun JD, Clark PU, He F, Marcott SA, Mix AC, Liu Z, Otto-Bliesner B, Schmittner A, Bard E (2012) Global warming preceded by increasing carbon dioxide concentrations during the last deglaciation. Nature 484(7392):49–54

Sheehan J, Dunahay T, Benemann J, Roessler P (1998) A look back at the US Department of Energy's Aquatic Species Program: biodiesel from algae, vol 328. National Renewable Energy Laboratory Golden, CO

Sheih IC, Fang TJ, Wu TK, Lin PH (2009) Anticancer and antioxidant activities of the peptide fraction from algae protein waste. J Agric Food Chem 58(2):1202–1207

Shirai F, Kunii K, Sato C, Teramoto Y, Mizuki E, Murao S, Nakayama S (1998) Cultivation of microalgae in the solution from the desalting process of soy sauce waste treatment and utilization of the algal biomass for ethanol fermentation. World J Microbiol Biotechnol 14(6):839–842

Sialve B, Bernet N, Bernard O (2009) Anaerobic digestion of microalgae as a necessary step to make microalgal biodiesel sustainable. Biotechnol Adv 27(4):409–416

Siddiq A, Dembitsky V (2008) Acetylenic anticancer agents. Anti-Cancer Agents Med Chem (Formerly Current Medicinal Chemistry-Anti-Cancer Agents) 8(2):132–170

Silva C, Fabiano LA, Cameron G, Seider WD (2012) Optimal design of an algae oil transesterification process. In: Symposium on process systems engineering, 2012, p 19

Singh A, Olsen SI (2011) A critical review of biochemical conversion, sustainability and life cycle assessment of algal biofuels. Appl Energy 88(10):3548–3555

Singh SK, Bansal A, Jha MK, Dey A (2011) Comparative studies on uptake of wastewater nutrients by immobilized cells of Chlorella minutissima and dairy waste isolated algae. Indian Chem Eng 53(4):211–219

Singh SK, Bansal A, Jha MK, Dey A (2012) An integrated approach to remove Cr(VI) using immobilized *Chlorella minutissima* grown in nutrient rich sewage wastewater. Bioresource Technol 104:257–265

Singh SK, Dixit K, Sundaram S (2013) Bioengineering of biochemical pathways for enhanced photobiological hydrogen production in algae and cyanobacteria. Int J Biotechnol Bioeng Res 4(5):511–518

Singh SK, Dixit K, Sundaram S (2014a) Algal-based CO_2 sequestration technology and global scenario of carbon credit market: a review. Am J Eng Res 3(4):35–37

Singh SK, Dixit K, Sundaram S (2014c) Effect of acidic and basic pretreatment of wild algal biomass on Cr (VI) biosorption. J Env Sci Toxicol Food Technol 8(5):38–41

Singh AK, Sad K, Singh SK, Shivaji S (2014b) Regulation of gene expression at low temperature: role of cold-inducible promoters. Microbiology 160(7):1291–1296

Sivaprakash B, Rajamohan N, Sadhik AM (2010) Batch and column sorption of heavy metal from aqueous solution using a marine alga *Sargassum tenerrimum*. Inter J Chem Tech Res 2(1):155–162

Skjanes K, Lindblad P, Muller J (2007) BioCO2: a multidisciplinary, biological approach using solar energy to capture CO_2 while producing H_2 and high value products. Biomol Eng 24(4):405–413

Slade R, Bauen A (2013) Micro-algae cultivation for biofuels: cost, energy balance, environmental impacts and future prospects. Biomass Bioenergy 53:29–38

Soletto D, Binaghi L, Ferrari L, Lodi A, Carvalho JCM, Zilli M, Converti A (2008) Effects of carbon dioxide feeding rate and light intensity on the fed-batch pulse-feeding cultivation of *Spirulina platensis* in helical photobioreactor. Biochem Eng J 39(2):369–375

Solovchenko A, Khozin-Goldberg I (2013) High-CO$_2$ tolerance in microalgae: possible mechanisms and implications for biotechnology and bioremediation. Biotechnol Lett 35(11):1745–1752

Sreekrishnan TR, Kohli S, Rana V (2004) Enhancement of biogas production from solid substrates using different techniques: a review. Bioresour Technol 95(1):1–10

Su F, Lu C, Cnen W, Bai H, Hwang JF (2009) Capture of CO$_2$ from flue gas via multiwalled carbon nanotubes. Sci Total Environ 407(8):3017–3023

Su H-Y, Lee T-M, Huang Y-L, Chou S-H, Wang J-B, Lin L-F, Chow T-J (2011) Increased cellulose production by heterologous expression of cellulose synthase genes in a filamentous heterocystous cyanobacterium with a modification in photosynthesis performance and growth ability. Botanical Studies 52(3):265–275

Suali E, Sarbatly R (2012) Conversion of microalgae to biofuel. Renew Sustain Energy Rev 16(6):4316–4342

Sydney EB, Sturm W, de Carvalho JC, Thomaz-Soccol V, Larroche C, Pandey A, Soccol CR (2010) Potential carbon dioxide fixation by industrially important microalgae. Bioresour Technol 101(15):5892–5896

Tang D, Han W, Li P, Miao X, Zhong J (2011) CO$_2$ biofixation and fatty acid composition of *Scenedesmus obliquus* and *Chlorella pyrenoidosa* in response to different CO$_2$ levels. Bioresour Technol 102(3):3071–3076

Tang G, Suter PM (2011) Vitamin A, nutrition, and health values of algae: *Spirulina, Chlorella,* and *Dunaliella*. J Pharm Nutr Sci 1:111–118

Tanger P, Field JL, Jahn CE, DeFoort MW, Leach JE (2013) Biomass for thermochemical conversion: targets and challenges. Front Plant Sci 4:218. doi:10.3389/fpls.2013.00218

Tiwari DN, Ashok Kumar, Mishra AK (1991) Use of cyanobacterial diazotrophic technology in rice agriculture. Appl Biochem Biotech 28(1):387–396

Toledo-Cervantes A, Morales M, Novelo E, Revah S (2013) Carbon dioxide fixation and lipid storage by *Scenedesmus obtusiusculus*. Bioresour Technol 130:652–658

Tsuzuki M, Ohnuma E, Sato N, Takaku T, Kawaguchi A (1990) Effects of CO$_2$ concentration during growth on fatty acid composition in microalgae. Plant Physiol 93(3):851–856

Tuzen M, Sarı A, Mendil D, Uluozlu OD, Karaman I, Soylak M (2009) Characterization of biosorption process of As(III) on green algae *Ulothrix cylindricum*. J Hazard Mater 165(1–3):566–572

Ueno Y, Kurano N, Miyachi S (1999) Purification and characterization of hydrogenase from the marine green alga, *Chlorococcum littorale*. FEBS Lett 443(2):144–148

Um BH, Kim YS (2009) Review: a chance for Korea to advance algal-biodiesel technology. J Ind Eng Chem 15(1):1–7

Vaishampayan A, Sinha RP, Hader DP, Dey T, Gupta AK, Bhan U, Rao AL (2001) Cyanobacterial biofertilizers in rice agriculture. Bot Rev 67(4):453–516

Van Den Hende S, Vervaeren H, Boon N (2012) Flue gas compounds and microalgae: (Bio-) chemical interactions leading to biotechnological opportunities. Biotechnol Adv 30(6):1405–1424

Vanegas CH, Bartlett J (2013) Green energy from marine algae: biogas production and composition from the anaerobic digestion of Irish seaweed species. Environ Technol 34(15):2277–2283

Vazhappilly R, Chen F (1998) Eicosapentaenoic acid and docosahexaenoic acid production potential of microalgae and their heterotrophic growth. J Am Oil Chem Soc 75(3):393–397

Vogel M, Günther A, Rossberg A, Li B, Bernhard G, Raff J (2010) Biosorption of U(VI) by the green algae *Chlorella vulgaris* in dependence of pH value and cell activity. Sci Total Env 409(2):384–395

Vonshak A (ed) (1997) Spirulina platensis arthrospira: physiology, cell-biology and biotechnology. CRC Press

Wackernagel M, Rees W (1998) Our ecological footprint: reducing human impact on the earth. New Society Publishers, Gabriola Island

Walker LA, Wang T, Xin H, Dolde D (2012) Supplementation of laying-hen feed with palm tocos and algae astaxanthin for egg yolk nutrient enrichment. J Agric Food Chem 60(8):1989–1999

Wang KS, Chai TJ (1994) Reduction in omega-3 fatty acids by UV-B irradiation in microalgae. J Appl Phycol 6(4):415–422

Wang X, Piao S, Ciais P, Friedlingstein P, Myneni RB, Cox P, Heimann M, Miller J, Peng S, Wang T (2014) A two-fold increase of carbon cycle sensitivity to tropical temperature variations. Nature. doi:10.1038/nature12915

Weaver BJ, Holob BJ (1987) Health effects and metabolism of dietary eicosapentaenoic acid. Progress Food Nutr Sci 12(2):111–150

Winkler M, Heil B, Heil B, Happe T (2002) Isolation and molecular characterization of the [Fe]-hydrogenase from the unicellular green alga *Chlorella fusca*. Biochim Biophys Acta (BBA) 1576(3):330–334

Yan D, Wang D, Yang L (2007) Long-term effect of chemical fertilizer, straw, and manure on labile organic matter fractions in a paddy soil. Biol Fertil Soils 44(1):93–101

Yan F, Chen Z, Li W, Cao X, Xue S, Zhang W (2011) Purification and characterization of a hydrogenase from the marine green alga *Tetraselmis subcordiformis*. Process Biochem 46(5):1212–1215

Yen H-W, Brune DE (2007) Anaerobic co-digestion of algal sludge and waste paper to produce methane. Bioresour Technol 98(1):130–134

Yongmanitchai W, Ward OP (1991) Growth of and omega-3 fatty acid production by *Phaeodactylum tricornutum* under different culture conditions. Appl Env Microbiol 57(2):419–425

Yoo C, Choi G-G, Kim S-C, Oh H-M (2013) *Ettlia* sp. YC001 showing high growth rate and lipid content under high CO_2. Bioresour Technol 127:482–488

Yoo C, Jun S-Y, Lee J-Y, Ahn C-Y, Oh H-M (2010) Selection of microalgae for lipid production under high levels carbon dioxide. Bioresour Technol 101(1):S71–S74

Yoon JH, Sim SJ, Kim M-S, Park TH (2002) High cell density culture of *Anabaena variabilis* using repeated injections of carbon dioxide for the production of hydrogen. Int J Hydrogen Energy 27(11):1265–1270

Yun YS, Lee SB, Park JM, Lee CI, Yang JW (1997) Carbon dioxide fixation by algal cultivation using wastewater nutrients. J Chem Tech Biotech 69(4):451–455

Yuan JP, Peng J, Yin K, Wang JH (2011) Potential health-promoting effects of astaxanthin: a high-value carotenoid mostly from microalgae. Mol Nutr Food Res 55(1):150–165

Yun YS, Lee SB, Park JM, Lee CI, Yang JW (1997) Carbon dioxide fixation by algal cultivation using wastewater nutrients. J Chem Technol Biotechnol 69(4):451–455

Zhang F, Endo T, Kitagawa R, Kabeya H, Hirotsu T (2000a) Synthesis and characterization of a novel blend of polypropylene with Chlorella. J Mater Chem 10(12):2666–2672

Zhang F, Kabeya H, Kitagawa R, Hirotsu T, Yamashita M, Otsuki T (2000b) An exploratory research of PVC-Chlorella composite material (PCCM) as effective utilization of Chlorella biologically fixing CO_2. J Mater Sci 35(10):2603–2609

Zhang L, Happe T, Melis A (2002) Biochemical and morphological characterization of sulfur-deprived and H_2-producing *Chlamydomonas reinhardtii* (green alga). Planta 214(4):552–561